YOUR PLACE IN THE UNIVERSE

YOUR
PLACE
IN THE
UNIVERSE

UNDERSTANDING
OUR BIG, MESSY
EXISTENCE

PAUL M. SUTTER

Prometheus Books

59 John Glenn Drive
Amherst, New York 14228

Published 2018 by Prometheus Books

Your Place in the Universe: Understanding Our Big, Messy Existence. Copyright © 2018 by Paul M. Sutter. All rights reserved. No part of this publication may be reproduced, stored in a retrieval system, or transmitted in any form or by any means, digital, electronic, mechanical, photocopying, recording, or otherwise, or conveyed via the internet or a website without prior written permission of the publisher, except in the case of brief quotations embodied in critical articles and reviews.

Cover image © Shutterstock
Cover design by Nicole Sommer-Lecht
Cover design © Prometheus Books

Trademarked names appear throughout this book. Prometheus Books recognizes all registered trademarks, trademarks, and service marks mentioned in the text.

Inquiries should be addressed to
Prometheus Books
59 John Glenn Drive
Amherst, New York 14228
VOICE: 716–691–0133 • FAX: 716–691–0137
WWW.PROMETHEUSBOOKS.COM

22 21 20 19 18   5 4 3 2 1

Library of Congress Cataloging-in-Publication Data

Names: Sutter, Paul M., 1982- author.
Title: Your place in the universe : understanding our big, messy existence / by Paul M. Sutter.
Description: Amherst, New York : Prometheus Books, 2018. | Includes bibliographical references and index.
Identifiers: LCCN 2018023961 (print) | LCCN 2018033313 (ebook) | ISBN 9781633884731 (ebook) | ISBN 9781633884724 (hardcover)
Subjects: LCSH: Cosmology—Popular works. | Astronomy—Popular works.
Classification: LCC QB982 (ebook) | LCC QB982 .S88 2018 (print) | DDC 523.1—dc23
LC record available at https://lccn.loc.gov/2018023961

Printed in the United States of America

For Dad.
Sorry I couldn't build a warp drive to Mars.
This will have to do.

CONTENTS

PERHAPS THIS WAS ALL
A BIG MISUNDERSTANDING

H ow the heck do you write a book about the whole entire universe? And
not just all the gory physics on scales small and great, but how our
knowledge of the cosmos has changed in the past few hundred years, and how
that's influenced our views of the heavens, the Earth, and ourselves? *And* how
we got to know what we know now, through all the twists and turns and dead
ends and blind alleys and just-kiddings of scientific research?

I honestly have no idea, so I suppose we're about to find out together.

If you're already familiar with at least some aspects of the cosmic tapestry
that I'm about to unfurl (unfold? I'm unsure here of fabric storage techniques),
then I sincerely hope you appreciate the slightly warped perspective I have
on the history of cosmology and the history of the universe. When it comes
to the past four hundred years, and the past 13.8 billion years, certain stories,
certain people, and certain physics have always captured my attention more
than others, and naturally I wrote about those and pretended the uninteresting
stuff doesn't matter.

If you're completely new to matters cosmological, well, then you're in for
a real treat. You're going to encounter some interesting (to put it mildly) characters, crazy physical processes, and, of course, seriously intensive and possibly
therapeutic discussions on deeply enigmatic mysteries of the cosmos. I promise
I'm doing my best to hit the right level between blow-your-mind and hold-your-
hand. But I don't know your background, your interests, or when you dropped
out of school, so don't worry if a section or two (or heck, the entire book) gets a
little confusing. Go ahead and give it another shot—I won't mind.

Of course I need to toss in an obligatory thanks to a good fraction of the
human race. From the dedicated scientists (or protoscientists, in some cases),
both named and unnamed, who actually figured out all this *stuff*, to all the
people who have supported me, guided me, taught me, told me I was wrong

(that happened a lot), and generally helped make me, me and this book, this book. You know who you are—and thank you for buying this book out of a sense of obligation—so you'll understand why I won't bother listing all your names. My publisher set a word limit, after all.

I'm sure that in some way I owe you a deep and sincere apology after you read this book. If you're a fan of history, then my choices to ignore/simplify/disregard certain aspects of the complicated and intertwined nature of human lives and pursuits might irritate you. If you're a fan of physics, then my choices to ignore/simplify/disregard certain aspects of the complicated and intertwined nature of natural processes might irritate you. If you're a fan of formal writing and good grammar, then you probably haven't even made it this far.

Depending on your own personal belief and/or philosophical system, there's a really solid chance at some point you will read something that will deeply, terribly offend you, causing you to hurl the book at the nearest wall. It's cool, we all do it. I just hope you know that it's not my *intention* to offend you—either with my style or my substance—but to play a game of show-and-tell with the universe. This is the story of the cosmos as revealed by the tools of scientific inquiry, which have so far proven to be pretty awesome in that regard. I personally find the heavens above us deeply profound, awe-inspiring, and worthy of further study, and I hope the humble paragraphs you're about to encounter (a) do the universe justice and (b) spark a similar passion in you.

But besides being a story of how we look at the universe, this book is also a tale of how the universe looks back at us—about our *relationship* with the night sky and how we (mostly mistakenly) think it intersects with our daily lives. Some of you (not necessarily *you*, but somebody) might take that as a critique on a particular religious or philosophical or other nonscientific belief. OK, fine. That's not the point of the book; but I'm not your boss and I'm not going to tell you what to do.

Is this book important? Is it necessary? Is cosmology—the study of our entire universe—vital to the advance of human civilization? Well, are *you* important and necessary and vital? In the grand sense of things, probably not. But we still keep you around and take you out to dinner, don't we? The game of science isn't to make the world a better place—although some scientists are, thankfully, engaged in exactly that—it's to make a better understanding of the world. Science itself is a method, a tool, for studying nature. That tool can be applied in many cases, from what's causing your migraine (besides chapter 4) to the origins, history, and contents of the universe.

Trying to figure out how nature works, even in literally unreachable parts of our cosmos, is an end in itself. That's exactly the point: to learn more, because being curious and learning stuff is kind of fun, if learning stuff is your kind of fun. And it's my kind of fun, so that's why this book exists. I have loved learning all these facets of cosmological matters through my youth and professional career. Beyond that, I *owe* it to you. It's you—the taxpayer, the friend, the supporter—who makes science happen. It's you who keep the lights on and the streets clean and the accounts in order and the beans sprouting, enabling a small percentage of the population to follow a particularly odd passion for wrestling with nature on a daily basis.

This book is yours—in a certain sense, you own the collective knowledge summarized in these pages. You also literally own this book, unless you stole it, in which case shame on you.

The most important thing to remember, above all else, no matter your level of familiarity, your personal beliefs, the ease with which you get offended, or any other trait that might affect your perception of this work, is to purchase many copies of this book to distribute to your friends, family, coworkers, acquaintances, mail carriers, local fire department, students, teachers, priests/rabbis/imams, pets, strangers, landscape architects, interior designers, representatives of the local AFL-CIO chapter, and government leaders, and of course a backup copy for yourself.

Science is for sharing, people.

So let's get moving.

CHAPTER 1

SACRED GEOMETRY

O f *course* the Earth is at the center of the universe. Just look at the evidence. The sun, moon, stars, and planets all wheel around us in the great celestial dance; how are we possibly to make sense of that if we're not at the focal point? And have you seen the Earth? It's large and solid and made of rock—are you seriously suggesting it *moves*?

Look up in the sky. Birds. Clouds. The wind. All moving effortlessly, right? They are well removed from this immobile rock on which we stand, so it stands to reason that movement comes even more naturally to the stars and planets, even further removed from us.

Besides, things up there are just so *different*. Our lives here on Earth are dirty, chaotic, even sinful. But the graceful movements of the heavenly realm are something else. Motions so precise we can use them to set our calendars. Unfailing, the same stars appear on the horizon at the precise time they did last year, and the year before, and so on into the unwritten time of our ancestors.

Surely the laws of nature that we understand here on Earth don't apply up there. They have their own rules, their own sets of laws that govern their behavior. The heavens surround us but are separate from us.

Oh, what about those comets and meteors? Surely they're just strange atmospheric phenomena. Don't worry too much about those.

Can we really blame our forebears for thinking we are at the center? Up until a few hundred years ago, it was the simplest and most natural explanation of the available data. Not only that, it was supported by rational, coherent arguments. Our ancestors, as we do, used multiple paths to understand the world around them—evidence-based, faith-based, reason-based, math-based—all of which pointed in the same direction: out.

Our ancestors were no dummies. They were just as smart as we are today and perfectly capable of understanding the world around them. And astronomy was vitally important to their daily lives: when to plant crops, when to reap the harvest, when to start preparing for campaigns, when to celebrate holidays. Humans have been using calendars for millennia (at least!), and the natural, regular, repeatable, predictable movements of the heavens provided the perfect device.

I suppose I should mention that ancient peoples thought that the stars influenced our daily lives too—not just by proxy through the effects of the seasons, but literally determined our fate. It's a unique perspective that's missing from modern scientific perspectives (with good reason, and I'll get to that in a bit). But again, I have to stress that your extremely-great-grandparents regarded their horoscope with as much seriousness as they could muster.

The motions of the stars and planets were connected to the seasons. So even though the celestial realm obeys its own set of laws, it must surely be connected—somehow—to events here on Earth. And despite the messiness and chaotic nature of our home, there's a sense of some sort of hidden order and regularity behind our lives. There are obvious patterns in nature, so perhaps there are unobvious patterns as well, patterns that can only be teased out by careful observation and interpretation.

Thus, the astrologer: someone who carefully studied the great wheel of heavenly motions and inferred their implications for the Earthbound. Born during a particular month? That must be linked to your personality. Solar eclipse during your reign as emperor? Yikes, better clean up your act. Comet appears as you prepare for battle? The gods disfavor your enemy, and the time is right for attack.

One of our earliest records of the profession of astrologer (or, if you prefer, astronomer; the two terms weren't cleanly separated until relatively recently) comes to us from ancient China, right as they were at the edge of developing a writing system. The story goes that the two court astrologers of the emperor Chung K'ang failed to predict a solar eclipse in 2137 BCE. They were immediately beheaded.[1]

Yes, our ancestors took this sky-watching stuff *seriously*.

So it's no surprise that as the centuries progressed, more data from observations accumulated, and astronomers were able to make ever-more-sophisticated models of those motions so they could make better *astrological* predictions. What better way to accurately tell your fortunes, an entrepreneurial young astrologer might say to a prospective royal client, than with the most precise measurements and predictions available?

And the Ptolemaic system, developed by Ptolemy (hence the name) in the second century CE and fully established as everyone's favorite cosmological model by the sixteenth century in Europe, provided the most detailed astrological calculations possible. This model of the universe put the Earth at the center (of course!), with each celestial body assigned to its own crystal sphere. These spheres nested within each other like a set of heavenly matryoshka dolls, gliding effortlessly against each other in their cosmic dance.

The moon's sphere came first, followed by one carrying the sun and one for each of the planets (except Uranus and Neptune, which were too dim to be known), with the outermost layer carrying the sphere of stars. Beyond that was probably heaven itself, or something like it.

It's a little difficult for our modern minds to wrap themselves around pre-scientific cosmologies. Individual statements or expressions sound perfectly normal in isolation. Even today you could hear someone talking about, say, the time of the next lunar eclipse, or someone lecturing about the nature of the divine. But today these kinds of statements tend to be widely separated. Nobody (who wants to be taken seriously) claims that the pattern of lunar eclipses gives us a window into Holy Wisdom and a clue about what we ought to have for dinner this week.

It's not that modern scientists are incapable of religious thought, but they usually don't think about both subjects at the same time, and it's rare for a scientific treatise to use religious texts to bolster its argument (and vice versa).

This compartmentalization of inquiry into the world around us is a relatively recent invention. For almost all of human history, people who were curious about the universe were simply that: curious about the universe. And one could inquire about the universe in many different ways: using evidence, using divine revelation, using rational arguments, using mathematical proofs, and so on.

So the highly sophisticated cosmological Ptolemaic model wasn't just a *physical* model of the order of the universe—it was fully incorporated into the religious, philosophical, and mathematical views of the time and place.[2]

I'm highlighting this blending of modes of inquiry into the natural world because I think there's something missing from the usual story of the birth of the scientific revolution. That story, put very simply, goes like this: we used to think the Earth was at the center of the universe, but that model was flawed. Copernicus proposed that the sun was at the center, Kepler refined this theory, and Galileo got in a big argument with the Catholic Church about it. Lots of fighting and a good amount of burning at the stake ensued, but eventually science prevailed, and now we know better.

I don't think that gives the right flavor of what went down at the turn of the 1600s. Don't get me wrong: Galileo fought with the church (a lot), people got burned (a lot), and a sun-centered model was adopted (eventually). But the impression that I, at least, got as I was taught this way back in grade school was that the arrogant know-it-alls thought the Earth was at the center, taking pride of place, and refused to accept the new view.

To be fair, most people of historic eras simply didn't think about this at all. They were too busy dying of plague, dying from starvation, or dying in battle to wonder about the precise mathematical formula that would unlock the inner workings of the celestial spheres. The arguments that have passed down to us are from the intelligentsia of the age: those who could read and write (usually in Latin), who had access to the books written by their intellectual ancestors, who had enough time to make precise measurements of the objects in the sky, and who had the ear of a king or pope or other wealthy dude to fund their studies.

So I can't tell you what Mathias or Marta Everyman thought about the universe, but I can tell you what Nicolaus Copernicus, Tycho Brahe, Johannes Kepler, and Galileo Galilei thought.

By the late 1500s, folks pretty much knew there were some issues with the Ptolemaic model of cosmology. The trouble, as is usually the trouble when theories begin to crumble, was data. Assuming that the planets move in perfect circles doesn't *quite* fit the observations. Sometimes the planets appear to move backward in their orbits, which just shouldn't happen.

The easiest solution is to add a so-called epicycle, a circle-within-a-circle to account for the extra motion. OK, fair enough. Circles are nice and elegant, and a series of nested circles within each orbit is only one step away from a series of nested spheres to explain the orbits themselves. There's a certain appealing symmetry there.

But in the centuries leading up to 1600, astronomers started making more accurate observations, and they noted deviations from the circles-within-circles approach. To explain this, they added ever more epicycles, each one tacked onto a particular planet to match all the observations.

It was a bit complicated, but it *worked*: it fit the data and was able to make predictions. You could tell your fortunes with the epicycle Earth-centered system. Maybe it was a bit cumbersome, so naturally you would need a few years of training before you could become an adept astrologer. But there's nothing wrong with baking a little job security into the system, right?

Besides, in some academic circles (ha!) the argument went that the epicycles were just added to account for observations. There was a deeper, more hidden "truth" in the universe, and cosmological models were just that: models to explain the data—nothing more, nothing less. A tool, if you will, that could be wielded to give better astrological predictions.

So when Nicolaus Copernicus on his deathbed in 1543 allowed his book detailing a sun-centered cosmology to be published, the reaction was pretty mild. "Hmmm" seems to have been the collective response of the leading figures. Some thought it was kind of nifty, others were violently opposed, but most folks simply didn't care. As the decades passed, however, the debates grew more complex, more heated, and overall more fun for us in the present day to read. And to be perfectly honest, the debates were, well, honest. Sure, some so-called enlightened figures had reflexive knee-jerk reactions, but most scholars armed themselves with the tools of the thinking trade—evidence, reason, philosophy, mathematics, divinity—and went to work trying to devastate their opponents. If you go to a typical scientific conference today, you'll see that some things never change. The weapons may be different, of course, but the modes of delivery are the same.[3]

Copernicus's fancy new model wasn't immediately compelling. He argued that—hey, guys, check this out, I know it's a crazy idea but work with me—the sun is the center of the universe, not the Earth. Look at the problems it solves! Sometimes planets move backward in their orbits? It's because we're catching up in their orbit. And you know some of those awkward pain-in-the-neck mathematical contrivances in the geocentric model, like epicycles? Well, you can safely chuck most of them if the sun is at the center. And—well, those are the big ones.

What, you're not convinced?

You're not alone. To account for the day/night cycle, we now have to claim that the Earth is spinning. Are you *crazy*? Have you seen the Earth? Wouldn't we be blasted by bajillion-mile-per-hour winds and/or spun off the planet? And Copernicus still insisted on circular orbits because circles are really beautiful, and so he still had to add epicycles to account for the detailed motions. (Although, to be fair to Copernicus, his system was mathematically simpler.)

And let's really think about this, OK? Let's assume, for the sake of argument so I can later prove you wrong, that the Earth orbits the sun. Wouldn't the stars wobble a little bit between summer and winter, based on our different observing positions, the same way our view off a distant object can wobble if we switch eyes? They don't, at least as far as we can tell. So either (a) the stars are unfathomably far away and our universe is way too large to comfortably think about, or (b) the Earth is at the center.

We're going to reject option (a) because it's immediately eye-rollingly wrong, so we're left with an Earth-centered universe.

That was the argument made by Tycho Brahe, the Danish astronomer who was really fond of making arguments: he lost his nose during a duel with his third cousin. Its brass replacement served as a useful warning to every other academic later in his life: don't flipping mess with Brahe.[4]

It also didn't hurt that Brahe was perhaps the foremost astronomer to ever appear on our planet. Working from Uraniborg, his own private fortress of science, his observations practically defined *exquisite*. He spotted a supernova. He figured out that comets were not, after all, merely some atmospheric phenomena. He crafted his very own personal cosmological model, with the Earth at the center, the sun orbiting the Earth, and everything else orbiting the orbiting sun (it wasn't very popular—I get the image of that one loud drunk guy at the party telling everyone his take on, say, the JFK assassination). And he collected the most detailed observations that had ever been made of the positions of the planets. All of this was done without the aid of a telescope—just looking and measuring, measuring and looking, night after night. He jealously guarded his tables of astronomical insights, allowing his assistants access only in heavily supervised scenarios.

Now I'm not saying that one of his assistants, Johannes Kepler, murdered him to gain unrestricted access to those tables. But it is awfully convenient that Kepler, who fervently believed (and I'm using "believe" here in a rather faith-based sense) that the sun was the center of the universe, was working with Brahe and desperately wanted unchaperoned access to those tables to prove his ideas right.

The story goes that Brahe was drinking heavily at a banquet and had to use the little astronomers' room but didn't want to get up for fear of insulting his host. In the ensuring days, he ended up very slowly and painfully dying, likely from a ruptured bladder.[5]

We'll leave that as the official line, but I'm keeping my eye on you, Kepler.

Johannes was a pretty clever dude, but saying he was eccentric is to only scratch the surface. He was the court astrologer (yes, you read that right) to Emperor Rudolph II of the Holy Roman Empire. That's not too surprising, but he was also a *numerologist* of the highest order.[6]

Kepler fervently—and perhaps fanatically—believed that mathematical and geometrical coincidences found in nature were anything but mere bits of chance. No. There was hidden order and deeper meaning within the motions of heavenly bodies. What's more, that order and meaning didn't just have consequences in the celestial realm but directly affected, influenced, and informed our daily lives. Right here on dirty, muddy, sinful Earth.

Perhaps there was a reason that although Kepler and Galileo wrote to each other, Galileo never really referenced any of Kepler's astronomical work in his arguments with the church.[7] You can kind of see why. To employ one of the arguments of Kepler would immediately open Galileo to accusations of being a—shudder—*mystic*.

But Kepler opened up that can of worms with relish and dug right in with the nearest fork he could find, drooling the whole time. He figured the sun was at the center of the solar system. Why? Because as the most prominent, fiery denizen of the solar system, it was *obviously* the focal point. Just like God the Father was the focal point of the Christian faith, the source of everything else.

His words, not mine.[8]

As nutty as Kepler was to us, he was no dummy either. He fully knew the weaknesses of Copernicus's solar system, but he thought he could do better. No—he *knew* he could do better.

And he had the data to do it. Table after repetitive table of positions of planets and stars, measured with as much accuracy as the human eye could muster. All his, now that Brahe was conveniently out of the picture.

Kepler was convinced that buried within those tables of numbers was a hidden order, and if he sought long enough he would recognize a pattern. This is, of course, before computerized pattern-matching algorithms, before computers themselves. A fancy mathematical tool called logarithms had just been invented, which was pretty handy, but otherwise Kepler had to brute-force the whole thing.

In his searching he found dozens of what we (and, to be frank, pretty much any rational person) would call coincidences. But to Kepler they weren't just random chance alignments or interesting repetitions of numbers. No, they were the voice of the divine, calling out through the cosmos, instructing us on how to operate our lives.

Kepler wrote a few books on the subject, and the vast majority of his ideas are nowadays ignored. What survives are what we know as Kepler's laws of planetary motion, which he discovered pretty much from trial and error, hoping to find *something* that stuck and unified the complex, interwoven motions of heavenly objects.[9]

The first law is that planets move in ellipses. Perfect circles for the motions of the planets just weren't cutting it anymore, even with a sun-centered universe. And besides, epicycles upon epicycles made horoscopes way too difficult to calculate—just where *exactly* is Mercury supposed to be on the day of the princess's wedding? Sheesh!

Kepler went about trying every geometric pattern he could think of, trying to find a common unifying theme to the planetary orbits. Apparently he initially skipped over the humble ellipse (which would be the first thing you would think of if you wanted to upgrade from a circle), assuming someone in the past millennium would have tried it already. He was wrong, and when he finally gave ellipses a shot, everything snapped together.

There it was, the voice of God himself speaking not through circles, but through ellipses. A simple geometric expression, combined with the placement of the sun at the center, put *all* the planets in their correct places. Nested epicycles with their complicated, convoluted mathematical machinery could be tossed into the garbage.

But circles are so beautiful! Surely the creator of the universe, in all his divine wisdom, wouldn't make a *mistake* like placing planets on elliptical orbits, right?

Kepler was ready for the criticism (I told you he was smart). You see, circles are a little *too* simple. A circle is a circle is a circle. You just need a single number, the radius, and you've completely defined it. But with ellipses you need *two* numbers: the major and minor radii, meaning the lengths of the long and short sides Two numbers means there's more bandwidth for God to tell us interesting things about the cosmos—and us.

And here's where Kepler's second law comes into play: the planets, as they swing around in their orbits, carve out equal areas in equal time. Think about trying to divvy up an odd-shaped pizza. You want everyone to get the same amount of pizza, so some folks will get a long and skinny slice, while others will get a short and wide slice. We don't need to get into the crust debate for the purposes of this analogy.

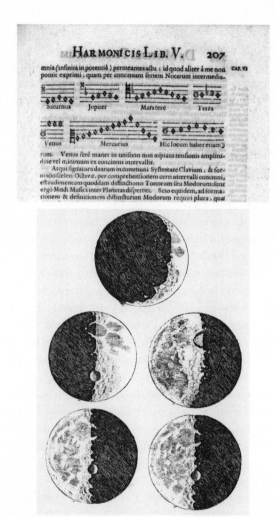

Understanding and confusion I: At the top, a short excerpt from the wild ride of Kepler's multivolume *Harmonices Mundi*, where he discovers Deep and Important truths about the celestial realm and uses those to link the motions of the planets to musical notes and scales and then to the fortunes of our daily lives. That last bit is arguable. Below, his contemporary Galileo Galilei scans the same heavens not with math but with a telescope and stares in blubbering wonder at what his polished glass reveals. In this case, in *Sidereus Nuncius* he sketches the rough-and-tumble surface of the moon, a far cry from the smooth marble finish it appears to be from afar.

An ellipse has two centers, called foci, and the sun is placed at one of those foci for each planetary orbit. When that planet is closest to the sun, in a given amount of time it will carve out a short and wide slice of orbit. Likewise, when it's farther away, it gets long and skinny pieces in the same timeframe. To accomplish this, the planet must move faster when it's closer to the sun, and slower when it's farther away.

Now for Kepler's big trick: the speeds of the planets in their orbit are telling us something. Or rather, they are singing us something. Kepler saw in the heavens the "music of the spheres," a celestial symphony singing their notes. And since the planets were closer to heaven, their song was truly a holy hymn.

Again, his words. Not mine.[10]

Kepler was particularly interested in the ratios of the slowest to fastest speeds, because to his eyes they looked very much like the ratios of notes used to make musical tones. But he didn't stop at the "Hey, that's neat" stage; instead he went all the way for the astrological touchdown and argued that the qualities of the notes played by the celestial denizens determined their character.

Most important, the ratio of the Earth's own speed was nearly 16:15 (and hey, if we have to fudge the numbers a bit to get a nice ratio, I'm sure nobody will notice), which was the same ratio as between the notes *mi* and *fa*. This was obviously right to Kepler, since "in this our home *mi*sery and *fa*mine hold sway."

This is seriously Kepler's line of thinking, and what motivated him to develop what we consider fundamental truths of our solar system. God created a perfect, harmonious universe, but we screwed it up with our sinful ways. So now only a few pockets of that primordial majesty remain: The planets, of course, since they're untouched by human affairs (ahem, at the time). Here on Earth we still have music, mathematics, and geometry, which, given the perfect relationships found within them, ought to have some glimmer of the divine.

To Kepler, the universe was permeated with a divine orderliness that was largely masked in our world but could be viewed in the heavens. So here was his easy-peasy divination horoscope-making plan: (1) Study the heavens. (2) Discover a hidden sacred geometry. (3) Relate it to a similar geometry on Earth. (4) Use that to tell the future.

Kepler went on for pages and pages of this kind of stuff, but there's only so much I can relate to you without going nuts, so we'll stop there.

Thus, Kepler saw the universe as messier than we had thought (having to jump from circles to spheres), but for a good reason: the sacred geometry of the sky was informing us of divine plans. But for several years one crucial element eluded him: something to unify the divine motions of the planets.

Finding this sacred geometry in the celestial realm wasn't enough. There had to be something deeper, even more fundamental. The elliptical orbits were useful, yes, but they only hinted at the heavenly. Kepler was hunting for a sort of *universality* in the laws that govern our universe, not just faint glimmers of connection. And it was by digging deeper into the details of planetary motions that he was able to uncover something truly astounding—to him, and to us even today.

Take Mercury, for example. It sits at a particular distance from the sun (or, I should say, particular distances, since its orbit is an ellipse), and it takes a certain amount of time to complete one of its orbits. The Earth has another distance and another amount of time, unique to this planet. And again, Jupiter, or Mars, or any planet, has its very own special pair of numbers assigned to it: some measure of its distance from the sun and some measure of how long it takes to complete one orbit.

These are all blindingly obvious and bland statements about our solar system that anyone could make. But have you ever wondered *why*? Why does Mercury have that number assigned to it, but not any other? Why does Jupiter take this long to orbit the sun, instead of slightly more or less time? It can't just be random coincidence, can it?

After years of searching through those tables of numbers, hand calculation after tedious hand calculation, Kepler found a formula that stuck. Today we know it as his third law, and I'm sure he considered it among his greatest achievements. The positions and speeds of the planets were not pulled out of some cosmic hat; there was indeed a deeper order.

Kepler discovered that the motions and positions of the planets obey a simple, harmonious relationship, and it goes like this: Pick a planet, any planet. Measure the amount of time it takes to go around the sun once. That's called the period, and you'll need to square that number. Next, draw the ellipse the planet makes in its orbit. Find the longest distance from the center of that ellipse to the edge. In math terms, that's called the semi-major axis. Cube that number.

The square of the period divided by the cube of the semi-major axis gives you a particular value. Repeat this exercise for all the planets and write down all their numbers—a tedious but pretty straightforward exercise.

And now for the voilà: that number is identical for all the planets, from tiny, fast Mercury to distant, slow Saturn. Across the solar system, this quantity is the same. It's an almost bizarrely simple formula unifying the motions of every planet. To us, this result seems almost pedestrian, but we're separated from this discovery by more than four hundred years of finding common elements among the stars. This was practically the philosopher's stone of astronomy. A holy grail of mathematical insight.

And almost nobody cared. Kepler was an eccentric mystic who buried this profound insight within volumes of—well, there aren't a lot of polite words to describe it. Still, as the decades wore on, scholars recognized the importance of Kepler's work, preserving it for future generations to ponder.

After all, his analysis only moved the goalposts; it didn't end the game. Sure, he could now handily explain the particular positions and speed of the planets. But why the period squared? Why the semi-major axis cubed? And this division thing, what's up with that? Kepler himself wasn't too concerned— the fact that he found any relationship at all was something to celebrate. As to what divine wisdom could be unlocked by understanding the source of the equation: well, it's hard to be employed as a mystic without some mystery in the universe, right?

Not that Kepler was entirely ignored—far from it. His head-scratching leaps of faith and logic were regarded with a sort of quaint curiosity, but behind the ramblings was the mind of a precise mathematician. He deftly employed his newfound organization of the cosmos—the elliptical orbits, the sun at a focal point, the underlying relationship among the planets—to great effect, publishing a set of tables that cataloged more than 1,500 star positions and predicted the positions of the planets to unheard-of accuracy, along with sets of calculation tools for handy do-it-yourself astronomy.[11]

That book, called *The Rudolphine Tables* in honor of his patron, was a smash hit (as much anything could be back then).[12] It was read, copied, and *used*, partly for navigation—accurate star positions are rather useful, after all—but mostly for astrology. Knowing which planet was precisely within which con-stellation was essential for understanding the present and divining the future.

It was that book, based on a more sun-centered model of the universe, that made all the difference. With the addition of elliptical orbits, everything was just so much easier. If making the sun the focal point of the cosmos made you uncomfortable, at least you could console yourself that it was only a model. The universe does what the universe does (and we can still think we're at the center); what Kepler did was introduce a handy trick of mathematics to make your astrological life so much easier.

So it's a bit ironic that Kepler is often included as some sort of patron saint of science. Sure, his achievements were remarkable. But so were Tycho Brahe's (and arguably, depending whether you lean more toward theory or experiment, Brahe's were superior), but unfortunately for old Brass Nose, he ended up on the wrong side of the argument. He did apparently have superhuman bladder control, but that didn't seem to help his scientific legacy.

Kepler was right—it is more accurate to think of the solar system as just that, *solar*—but for all the wrong reasons. And his legacy survived not on its scientific merits but on its astrological usefulness. Scholarly discourse slowly abandoned the concept of heavenly crystal spheres in the decades following Kepler's input. There simply wasn't a need: it was much easier to talk about the universe as if the sun hung at the center and the planets zoomed along on their little elliptical racetracks.

But his story serves as an object lesson for this book. The universe is far, far messier than we would prefer it to be. Wouldn't it be great if a few simple circles (and heck, let's be generous and toss in a few epicycles too) were enough to completely describe the cosmos?

Unfortunately, Mother Nature isn't that kind to us. Ellipses are more complicated than circles, for sure, but to Kepler that was a boon rather than a curse. With this much more rich information, he had enough clues to perceive an orderly pattern. He didn't fully understand the causes or physical implications of his third law, but he did discern it.

And that's why Kepler, as nutty as he was, even in the eyes of his peers, was onto something good.

Oh, right, Galileo. He was kind of a big deal, operating at the same time as Kepler, living in Italy in the shadow of his frenemy the Catholic Church. He was the astronomer's astronomer and the curmudgeon's curmudgeon. A rude dude who straight up called his boss an idiot (pro tip for the Renaissance scientist: don't tick off the pope, who also happens to wield supreme temporal power over your land). I'm not saying the church was in the right putting him under house arrest for arguing against the orthodoxy, but Galileo didn't exactly do himself any favors. The full story of Galileo is wonderful and convoluted and wonderfully convoluted, and also the subject of another book.[13]

Instead, I want to focus on what Galileo saw with his newfangled telescope. The instrument itself is deceptively simple. On paper it looks so easy a kindergartener could assemble it: a couple of curved lenses and a tube. The physics of optics does all the rest. And while Galileo didn't invent the telescope per se, he certainly invented the *astronomical* telescope, which is what we usually think of when we use the word "telescope."

The device had been developed within Galileo's lifetime, and, people being people, I'm sure somebody somewhere pointed it up at the night sky to see what he could see. Unfortunately, that person couldn't see much; the trick to telescopes isn't in their construction but in their finishing. You have to polish the lens to an extremely high level of precision so the path of light gets bent in just the right way, or you get a smudgy mess on the other end.

A telescope does two things. First, it's just a bucket for light. Your pupil can only cram so much light into it at a time, and that sets a limit to the dimmest thing you can see (I'm, uh, glossing over some biological details here, but you get the idea). A wider telescope is like a wider eyeball: it can soak up photons that would normally just hit the rest of your face.

The second thing a telescope does is magnify. It takes all that collected light and focuses it down to a smaller area so that it can fit in your eye. That operation turns small separations into big separations, so a minuscule dot on the horizon can be recognized as, say, a ship.

"Hey, guys, there's a ship on the horizon . . . I think" was the current gold standard for telescopes until Galileo took a crack at it. With uncommon dedication, he made a lens so smooth, so flawless, that the celestial realm completely changed its character.[14]

The moon, thought to be a smooth globe of marble, was instead a loose collection of jagged rocks. Jupiter was not alone—it was joined by four small moons that were obviously orbiting it. Saturn had two lumps on either side of it that, over the course of Galileo's repeated observations, shrank to thin slivers and disappeared—before reappearing again.

The sun had spots that, after a little geometrical legwork, were shown to be attached to its surface. And that surface was spinning. Fast.

Venus had phases. Phases. Like the moon. Since phases are a trick of perspective—the light from the sun illuminates only one side of an object at a time, and that side can be different from the one we view—the only way to make that work was for the sun to be at the center of the solar system.

The universe that Galileo revealed was frightfully messy. With this joining Brahe's earlier discovery of a new star appearing and his demonstration that comets were not confined to our atmosphere,[15] scholars (at this early stage, I hesitate to call them scientists in the sense we're used to) were quickly realizing that our home is a strange place indeed.

What got Galileo into trouble (well, one of the things) was his insistence on circular orbits. He saw, right through the lens of his telescope, direct evidence for a sun-centered cosmos. He must have been aware of Kepler's work—Johannes wrote to him like an eager fanboy—but he either missed the memo on elliptical orbits or outright ignored it because it was caught up in mystic mumbo jumbo.

It's interesting that Kepler and Galileo were tackling the same problem (the structure and contents of the universe) using different techniques (Kepler digging deep into mathematics and Galileo producing literal volumes of observations) and arriving at conclusions that seem at odds. But on closer inspection, their results were two sides of the same coin.

Galileo saw firsthand a messy universe but insisted on simple, orderly circles to explain the orbits of the planets. Kepler found elegant geometric order in the chaos but argued that the universe was directly connected to our everyday lives. I can't think of better prototypical experimentalists and theorists.

Put them together and you have the picture that modern cosmology—the study of the cosmos—brings us: we live in a simple universe that is connected to our daily lives.

Wait, no, that's not right. It's the other combination: we live in a messy universe that doesn't care about us.

That's it.

CHAPTER 2

A BROKEN UNIVERSE

Y ou know, it's really hard to start the story of our universe at the place it
should—the start of the universe—because strictly speaking, the universe doesn't *have* a beginning. Or maybe it does.

It's complicated.

Here's the problem. We live in an expanding universe. Every second of
every day, galaxies are generally getting farther away from each other. Run the
clock backward, and you quickly realize that in the past, galaxies were closer
together. This is pretty straightforward logic, I suppose.

Run the clock back further, and there aren't galaxies anymore—all the
matter is too smooshed together. There's just a bunch of junk filling up the
universe. Keep going, and eventually the entire universe, as big as it appears
today, shrinks and shrinks to an infinitely small point. That is precisely what
general relativity, our mathematical tool for understanding the evolution of
the universe over these almost incomprehensible timescales, tells us: that at a
finite time in the past (spoiler alert for later chapters: around 13.8 billion years
ago), the universe was compressed into a singularity, a point of infinite density.

Of course, that's wrong. Nature doesn't like infinities. But we have to separate what nature *actually does* from how we *model* it. Physics is a mathematical
description of the world around us. And that mathematics provides a wonderful tool. For one, math compacts vague natural-language sentences into
precise, meaning-filled equations. You could go on and on about how a wave
of electromagnetic radiation propagates (and in fact I'm probably going to
do that later) without getting all the details right, simply because English isn't
really equipped to describe it. But a couple of neat equations can do all the
work for us, summarizing and describing in one elegant go.[1]

Math also forces logical consistency, which is handy when a physicist wants
to make predictions. "If we assume *this* is true based on the evidence, then that
must happen" is a soothing statement that calms us during bouts of midnight
existential dread.

But it does have its shortcomings. We can only provide approximate descriptions of nature, because we are fundamentally limited by uncertainties in measurement. Sometimes those approximations are really really (really) good, so we don't care about the discrepancy anymore. But sometimes the math is so broken, so dysfunctional, that we can't use it to piece together what Mother Nature is whispering to us at all.

And that's the case with singularities. When a singularity appears in a description, it's the mathematics itself telling you that you've gone too far— your fancy equations are no good here. It means you're doing something wrong, and the math wants no part of your shenanigans.

Singularities are actually pretty common, at least in mathematics. If I model the force provided by a simple spring, I can use Hooke's law: The more I compress the spring, the harder it will push back on me. But if I were to compress it all the way, so that the spring was a single point, the force pushing back on me would be infinite. That seems like a dumb thing to say, but that's what Hooke's mathematics tells us.

But duh, you can only compress a spring just so far until some other force takes over—like, say, the electric repulsion of the atoms in the metal, preventing them from squeezing down to infinity. The singularity appears in the math, but nature knows better.

Now take this example and replace Hooke's law with general relativity, and the spring with the entire universe. Welcome to the cosmologist's nightmare.

Einstein's theory of general relativity is fantastically easy to state: we all live embedded in space-time. The presence of mass or energy distorts space-time "underneath" it (in a three-dimensional sense, but there aren't any good English words for it), and that distortion tells other matter how to move. Put two kids on a trampoline: they can affect the motion of each other without touching. Now imagine that in four dimensions, and you have gravity.

It's almost funny how a somewhat benign statement like the one above masks such a horrible snake pit of mathematics. General relativity itself is a set of ten complicated, interwoven equations describing the detailed relationship between space-time and matter and energy, plus some more equations governing how objects move. Heavy stuff.

General relativity is the story of gravity, and at the very largest scales (i.e., bigger than galaxies), gravity is the only force with enough oomph. It's incredibly weak but affects all matter and energy across infinite range, and it's that staying power that provides its dominance at large scales. On balance the universe is neutral, so the electromagnetic force cancels out. The strong nuclear force peters out past an atomic nucleus, and the weak nuclear force is, well, weak. When it comes to cosmology, only gravity matters.

Except when it doesn't, like in the earliest fraction of a fraction of second into the existence of the universe as we know it.

Cosmology is the story of gravity at large scales, and general relativity is just fine and dandy for describing that gravity—giving us the very history of our cosmos. But there are singularities lurking within Einstein's greatest hit. We use general relativity to understand the evolution of our universe, and those mathematics say—and logic tells us—if we live in an expanding universe (and we do!), then at one time in the past, everything was crammed into an infinitely dense point.

Unless something got in the way. Unless something prevented or replaced the inevitable collapse, stopping the universe from becoming infinitely dense—just significantly or even stupendously dense. Anything but infinite and we can handle it. And in fact, something *must* prevent it. We know that singularities don't actually appear in nature; they are flaws in our model of reality.

So general relativity draws a line in the sand: we could mine the entire universe for information, extract every single bit of useful data from our observations, but we will never, *ever* understand the earliest moments of the universe unless we try some other tool in our mathematical toolbox. When it comes to the first 10^{-43} seconds, we simply have no idea what's going.

It's here where speculation reigns. We have a relatively firm grasp of the history of the universe, but not its origins. Or even if *origins* is the right word. We haven't (yet) developed the mathematical language to describe these initial moments. Or even if *initial* is the right word.

Maybe there's more "universe" happening before what we call the big bang (as you can see, this term is meant to describe not the birth of our universe but its very early history). Maybe the very concepts of space and time break down. Maybe there's no such thing as "before" the universe. There are a lot of speculative ideas beyond our known physical laws floating around in the minds of physicists and their arcane academic journals. Maybe string theory has an answer here, or was it loop quantum gravity? Who knows?[2]

The singularity draws the eye of the theoretical physicist because it's an unambiguous signpost that there's more to be learned here. For all other times and scales in the universe, we have at least some guidance on how to proceed. But in this moment, when the universe has a temperature of more than 10^{32} Kelvin and is no bigger than 10^{-35} meters across, we don't have much to guide us.

To give you a sense of the extremity of the tininess of this epoch—called the Planck era, by the way—hold out your hands as far as they can reach. Go ahead, you could use the stretch. The scale difference between the width of the observable universe *today* (about ninety-three billion light-years) and your outstretched hands is about the same as the scale difference between your outstretched hands and the size of the universe at this newborn state—if the universe were about a billion times wider back then.

In other words, imagine how long it would take to stretch out your arms so wide that you could hug the present-day universe. The Planck-era universe would have to do that a billion times over to give *you* a friendly squeeze.

I'm sorry—there just aren't a lot of useful analogies for these sorts of situations. The temperatures, densities, and length scales in operation at the early stages of our universe aren't something we normally come in contact with, so we don't have handy metaphors or explanatory tools at the ready. Usually in these situations we can instead appeal to the mathematics to guide us, like a distant lighthouse in a storm of nonsense, but even here the light is gone out.

We don't understand the Planck era because we don't understand physics at those scales, but it is at least reassuring to know where our laws break down. And that Planck era, with its characteristic energy and length scales, isn't just pulled out of the Random Number Jar to sound impressive—there's a reason our understanding goes so haywire at numbers so small.

Let's say you have to pick your top five most important physical constants that appear in our descriptions of the universe. Actually, skip that; I'll just tell you what they are. You have the familiar speed of light, which appears in electromagnetism and special relativity, and you need to include Newton's classic gravitational constant. You might not recognize the Coulomb constant, which is like the gravitational constant but for electricity, but it's important too. Plus you need to toss in the Boltzmann constant, which for gases connects things we can easily measure (like temperature) to things we can't (like energy of the gas particles).

Last but certainly not least is the Planck constant itself, warily introduced by Max Planck in the early 1900s.[3] Originally shoehorned into his math to help him understand the behavior of light (and one of its uses is to define the relationship between the wavelength of light and the amount of energy it carries), it quickly grew to become the numerical ambassador for quantum mechanics itself. I'll get into that story more later.

All these constants help us relate one thing to another. If I have this much charge, how strong will its electric force be? Coulomb can tell us. If I start waving around some electric and magnetic fields, how fast will they move? Speed of light. How much mass do I need to hold down my coffee? Newton's got your back.

These constants are like little governors for their respective domains of physics. What's really fun is when we start *combining* physics, and the relationships between these constants tell us when and where multiple areas of physics get together and start partying. To find out, we combine the governing constants in interesting ways, known as the Planck units.

Take the Planck length, which combines Newton's constant, the speed of light, and the original Planck constant (sorry that his name keeps popping up in confusing ways; he was kind of smart and kind of important). That number comes out to about 1.6×10^{-35} meters . . . hey, wait a minute, that's the size of the universe where our physical theories break down!

Now we can see why: the Planck constant tells us about where we need to care about quantum mechanics, and Newton's constant tells us the same for gravity. The Planck length is a scale where both gravity and quantum mechanics matter. It's not like the universe magically transforms into something weird and wonderful at those scales, but since the Planck length is constructed from other useful numbers, when you're trying to play the gravity-quantum matchmaking game, those same numbers are going to crop up. So think of the Planck length as a warning sign: here be dragons.

And those are quantum gravity dragons. We do not, at the time of the writing of this book and most likely also at the time of your reading, have a quantum description of gravity. Why not? That's a long story, and probably a tale for another book.[4] For our purposes, we simply don't have a single mathematical language that incorporates gravity into our quantum worldview. And it's that lack of language that prevents us from fully understanding those first moments in the universe, and especially what might come "before."

I'm purposely not spending a lot of time discussing what some speculative ideas might be for combining gravity with quantum mechanics and for the origins of the universe. Things like colliding branes and collapsing universal wave functions sound super awesome, but at this point they're one step away from pure fantasy. Not that such hypotheticals shouldn't be explored—far from it. We need creative ideas to move forward. But we're so far out on the ledge that it's hard to tell a good idea from a bad one. I would hate to waste your precious time describing in detail a concept that a year from now will be ruled out by experiment (or, more likely, go out of fashion), especially when there's so much more cool stuff to talk about.

It's the ultimate irony that the work of Kepler would set us down a path of completely reimagining the universe and our place in it. He believed that the motions of heavenly objects contained divine truths that could be expressed in pure mathematics, and now we're starting our cosmological story—the very birth of the universe, if that's even the right turn of phrase—in murky haze that even our most sophisticated mathematics is only beginning to penetrate. Not only does the universe barely care for us, we don't even have the right language to begin caring about the early universe.

Once things get going, so to speak, the veil begins to lift. We're still not at the stage where we can directly observe anything (that won't come for another 380,000 years, an incomprehensibly distant future compared to the timescales of physics in the early cosmos), but we do begin to have hints of the mathematics involved.

Now I get to say one of my favorite phrases, a sentence that captures my own imagination, that summarizes our current state of existence as viewed through the lens of modern physics: we live in a broken universe. That trite statement is motivated by a simple question. Why are there *four* forces of nature? Electromagnetism, gravity, and the twin nuclear forces describe the complete variety of physical interactions in the world around us. Why not three? Why not seventeen?

Here's another amazing statement: there weren't always four forces of nature. Or rather, there aren't always four forces of nature.

When we crank up the energies in our particle colliders above 246 GeV (short for giga-electron volt, or a billion electron volts[5]), the four forces of nature disappear.

Gravity and strong nuclear are preserved, but the electromagnetic and weak merge into something else, handily called the electroweak force. The photon, carrier of the electromagnetic force, disappears. As do the W+, W-, and Z bosons, the inscrutable names we give to the carriers of the weak nuclear force. Their identities and everything we associate with them simply melt away. In their place, a quartet of new particles does the work of ferrying the electroweak force from place to place.

To make this splitting happen, Peter Higgs in the 1960s realized that there needs to be a new ingredient in the recipe of the universe: a Higgs field (so named later; he was not so vain as to name it after himself).[6] This field simply exists, permeating all of space-time, hanging out, minding its own business. But through *insert complicated mathematics here* it does the work of splitting the electromagnetic force from the weak nuclear. No Higgs, no symmetry breaking, no separation of forces. Hence all the buzz in 2012 when the Large Hadron Collider found evidence for the Higgs field.

The Higgs field is that one guy on the floor who finds a couple waltzing together perfectly, like they belong together for the rest of their lives, and asks for a dance.

That merging of the fundamental forces is an expression of a deep and fundamental symmetry in our universe. A single mathematical expression describes both the electromagnetic and weak nuclear interactions in one fell swoop—but only at high energies. At low energies (and our everyday experiences qualify as "low energies" in this context), that symmetry breaks, revealing two separate forces.

A pencil balanced on its tip is very symmetrical, but also unstable. It can only exist under a very narrow set of circumstances. As soon as it falls, it's much more stable, but asymmetric. Meditate on that the next time you flick the light switch.

It gets better, and where it gets better is relevant to the early universe. At even higher energies, somewhere north of 10^{16} GeV, the strong nuclear force joins the unification party, leaving only two forces at play in the fresh universe: gravity and what's called the electronuclear force, a Grand Unification of three forces in the universe. Obviously that unification is a bit harder to study because (a) the math gets hard, quick, and (b) that energy scale is a trillion times higher than what we can achieve with the world's most super of supercolliders.

Apparently that unification happened only once, for a brief moment, when the universe was less than 10^{-36} seconds old and smaller than a single electron. The words "brief" and "small" in the preceding sentence seem hopelessly inadequate, but here we are.

At the end of that Grand Unification epoch (also known as the GUT epoch, for this is the age when a supposed Grand Unified Theory prevails—but don't mistake a GUT for a TOE, a theory of everything that would hold in the Planck era) (no, I'm not making any of this up), the strong nuclear force splits off and we're back in (kind of) familiar territory with gravity, strong nuclear, and electroweak forces bouncing around. But in terms of narrative of the universe, that's still practically a lifetime away.

I don't want you to get the impression that we actually know what happens in this era. It's not as smoke-and-mirrors as times in the Planck scale, but it's very, very hazy stuff indeed. We do not yet have a single Grand Unified Theory that presents three of the four forces of nature in a coherent fashion. We have hints, we have clues, we have some scrappy-looking backwoods trails, but that's it.

At our best guess, as the universe expanded and cooled, at some point the energies lowered enough for the strong force to split away, delivering a solid GUT punch to the universe.

You'll see why that's more than just an amazing pun in a little bit.

If you've ever made ice cubes, you've witnessed one of nature's minor miracles: a phase transition. There's nothing different about the water molecules themselves in liquid versus solid, yet at different temperatures those molecules exhibit wildly different properties. The phase transition marks the boundary between ice and not-ice, between free-flowing liquid and rigid crystal structures. It's abrupt, too, which is what makes phase transitions so fascinating. It's not like the water slowly and gradually transforms into ice, taking its sweet time. As soon as a critical threshold is reached, *bam*, now you're looking at a block of ice.

Go ahead and reach into your freezer for an ice cube. You may notice internal cracks and fissures. Hmmm. Usually we associate cracks with trauma; something made that defect. But how could something cause such a crack deep within the ice, where before there was only pure water?

The reason is that ice does take a little bit of time to form, after all, and it doesn't form as soon as the temperature hits its special point. You can hold (not with your hands!) liquid water below its freezing point for as long as you want. It takes a nucleation point, a trigger, to start the ice-forming party. Once a couple of molecules decide to line up in nice, neat regimented order, they convince their nearby relatives to join in, followed by their surrounding close friends, then coworkers, then acquaintances, then fellow standers in line, and so on, until the entire liquid is transformed.

But what if more than one spot gets the same idea at the same time? There's nothing stopping a bowl of water from having multiple nucleation points. And each nucleation point will choose a particular random orientation for its ice arrangement. The water will crystallize moving outward from each independent point, with each realm having a different configuration.

You'll still get ice, but it will be multiple arrangements of ice smashed together. One region, seeded by a particular nucleation point, might be of the up-down variety, while a neighboring region, seeded by an independent-minded point, might be of the more left-right persuasion.

And where those domains meet, you get a defect. A flaw in the ice that marks the boundary between two hostile neighbors. There's nothing you can do about that flaw, either; it's literally frozen into place.

Now, enough about cold water—let's talk about the hot universe. As the early universe expanded and cooled, it underwent dramatic phase transitions. One transition occurred when gravity (hypothetically) split off from the (speculative) unified force. We know hardly anything about that process and how it might affect the later universe, so I'm just going to skip it.

The next phase transition occurred when the strong force splintered from the (still pretty hypothetical) electronuclear force. And just as in a phase transition with liquid water cooling into ice, defects appeared.

These defects are strange beasts indeed and come in a suitable menagerie of horrors: The magnetic monopole. The cosmic string. The domain wall. Monsters of crippled space-time and exotic forces; relics of the ancient universe.

By far the most common is the monopole. You may have noticed in your daily life a strange dichotomy between electric and magnetic forces: an electric charge can be either positive or negative, and the two can happily live quite separately, but magnets always come in pairs. Slice a north-south bar magnet in half, and you get two little baby north-south magnets in its place. This is due to the way that magnetic fields are produced in our normal everyday experience: not as isolated charges, but through the motions of charges.[7]

But the early universe is not within our normal everyday experience. It's thought, based on our current understanding of Grand Unified Theories, that when the strong force broke off, that phase transition led to defects, and one of those defects acted like a massive single pole of magnetism. A lonely north or a solitary south. A freak of nature.

Our simplest theories of what went down around the 10^{-36}-second mark in the age of the cosmos predict an absolute flooding of space with these creatures. Creatures that should be easy to spot even billions of years later. I mean, come on—they'd be pretty dang obvious, right? And yet we don't see a single one. No evidence of even one monopole floating around our vast cosmos.

So where did they go?

Our best answer, as first cooked up by physicist Alan Guth in the early 1980s, was that they were just—are you ready for this?—*inflated* away.[8] Blasted off to the four corners of the cosmos. Spread so far apart that they simply don't matter. I know, I know—hang in there with me.

Just as the end of the GUT era generated its own problems, it also provided the seeds of their resolution. Guth's hypothesis went like this: suppose, for instance, that for a fleeting moment the universe got really, really big, really, really fast.

Voilà: inflation.

This proposed solution to the monopole problem is brutal in its efficiency. Yes, many monopoles were produced, but the entire universe is simply much, much larger than our local observable patch. It's so much larger, in fact, that despite the monopolar infestation in the early cosmos, our universe is now so darn big that you only expect to find one monopole—tops—in the volume that we can actually see. No monopole, no problem.

This reasoning sounds kind of fishy, and it kind of is. But inflation theory has stuck around for the past few decades because (a) it's not as crazy as it sounds, and (b) it also solves a bunch of other problems in cosmology. Oh, and (c) we have evidence for it, but that comes later in our story. I promise.

The explanation for inflation lies in the very heart of the GUT transition itself. Just as later on the electroweak force will split apart with the handy help of the Higgs field, there ought to be another field around to do the work of prying the nuclear force away from the warm embrace of everybody else. While we certainly don't know the nature of that field, it *could* have certain beneficial properties. Like, say, being staggeringly repulsive.

That repulsiveness is *physical*, driving the expansion of the universe to rates never seen before and never to be seen again, taking a mere 10^{-32} seconds to inflate from the size of an electron to the size of a golf ball. That may not sound impressive, but when was the last time you saw something—anything— grow by a factor of a billion billion billion billion in a billionth of a billionth of a billionth of a billionth of a second?

That's what I thought.

Inflation is attractive because it solves (more like sweeps under the cosmic rug) the monopole problem, but that's not enough. After all, the monopole "problem" may not exist at all! It's an artifact of high-energy physics, a regime that we don't know too much about and honestly can't speak to with any level-headed confidence. So at first blush, inflation is using one speculation to cancel out another. The sum: we still have no idea what's going on.

That's a legitimate gripe, so I'll let it stand. But in defense of inflation, let me offer exhibit B: the horizon problem. And if you actually flipped to the back of the book to read the previous citation, you already knew this was coming. Look in any one direction of the universe. I mean *way* look, as deep as you can see. Measure the properties of the stuff you find there. Say, the average temperature of that patch of space, or the average distance between galaxies. Just, whatever.

Now pick a completely different direction and go as deep as you can there. Repeat the above exercise. You'll find, much to your delight and amusement, that the universe is very boring. At large enough scales, our cosmos is pretty much the same from place to place.

That by itself is not that big of a deal until you do a little math: the universe is about 13.8 billion years old and has a diameter of around 90 billion light-years. Yes, that means the universe expands faster than light (90 being a larger number than 13.8, after all), but no, that's not a problem, and yes, I'll talk about that later.

Right now we have a deeper problem. The only way for one patch of the universe to know what everybody else is up to is by interacting via some force—gravity and electromagnetism at these scales, specifically. And those forces take time to propagate, because, you know, the speed of light. So is it a totally random coincidence that at the largest scales, one patch of universe *just so happens* to look like any other patch? When did they exchange memos to start coordinating? When was the note passed around the cosmic classroom so everybody knew that in 13.8 billion years, we all need to look the same?

The answer: before inflation. When the universe was just a tiny little nugget, there was plenty of time for everyone to compare notes, settle into equilibrium, and agree on what their future course ought to look like. Then *swoosh*, inflation kicked in, sending bits of the universe zooming apart, never to talk again.

Direct communication is no longer possible from one distant corner of the universe to another—they're just too far apart, and only getting farther. But they still look and act relatively the same. The only way to resolve this paradox is inflation.

There's another problem that inflation handily solves—the flatness problem—but describing "flatness" belongs to another chapter, so we'll leave it at that.

But wait—there's more!

Inflation has one more trick up its quantum sleeve, One more reason to think that hey, maybe this isn't such a harebrained, cockeyed concept after all. And that concerns you and me. That's right: you. And me.

Imagine a ball rolling down a gentle slope to the bottom of a valley. It starts with a lot of potential energy, and as it rolls, it converts that potential energy into kinetic energy, speeding up in the process. Maybe there's a little friction as the ball rubs up against the grass. That's fine. Once the ball reaches the bottom of the valley, however, its story isn't over. It will wiggle around, sloshing playfully back and forth in the grass before finally settling down.

The grass underneath that ball will get rubbed, over and over, as the ball continues its wiggling. In the end, the ball transfers all of its energy into the irritated grass underneath it.

I want you to imagine the process of inflation—of the stupendous expansion of the early universe powered by an exotic quantum field—as a cute little ball slowly rolling down a hill in a sunlit meadow. When inflation is done, after that fraction of a second, the field has nearly expended nearly all its energy to swell the volume of the universe and is now settling into a comfortable retirement.

Inflation has done its job—it flung the monopoles as far apart as inhumanly possible—but it did its job a little too well. Not only did it separate any possible monopoles, it also separated *everything else*. If there was any matter at all in that hot second, it's been cooled and spread way too thin after the sudden expansion.

But like I said, inflation isn't done. Before the field responsible for driving the process completely settles down, it wiggles. Just a bit is all it takes. Strictly speaking, it *oscillates perturbatively within its potential*, but for us the word "wiggle" will suffice. And that wiggle breathes life into the newly cold universe.

According to our best understanding (which, I should admit, isn't all that best), the field driving inflation decays, transforming itself into a flood of particles, the fundamental particles that make up our everyday existence. That last-minute wiggling does all the work—in a final selfless gift to the universe, the *inflaton* (the vaguely cool-sounding name we give to the inflation field) does settle down into the background, never again to play a significant role in the rest of the history of the universe.

But its children inherit the universe that inflation shaped: one that is large, free of defects, empty, and ready for settling. Ready for a flood of particles, forces, and fields: the same characters that still inhabit the cosmos and are responsible for star formation, nuclear physics, magnetic fields, dodgeball, and life.

I know this chapter must be almost as frustrating to read as it was to write, what with all the *what-ifs* and *maybes* and *perhapses*, but that's the way it is. The universe is a fantastically complicated place, and as you can tell, that goes more so in its heady early days. But don't worry; we'll return again to the topic of inflation and reveal its true power—and some evidence.

So much happens in that first slice of an instant, and so much of that is inaccessible to us, both in observations and in theory. We don't have the gear or the mathematics to fully study this epoch of our universe. We're doing our best, though, so cut us some slack. Besides, at this point in the story, our entire universe is still only a few centimeters across and has a temperature of ten trillion billion degrees. And things are only going to get more interesting from here.

CHAPTER 3

TALES FROM A BEWILDERING SKY

O f course it was Sir Isaac Newton, the smartest person since himself, who figured out how Kepler's system works. For decades the question had been burning in the minds of academics: Ellipses certainly seemed the way to go—they were way too useful to be ignored—but *seriously*? Ellipses? How the heck do we explain that?

From Kepler's own work and additional Deep Thoughts, scientists (or, at least, protoscientists) realized the sun must exert some sort of cosmic influence on the planets. The concept of nested crystal spheres, so *en vogue* centuries earlier, was simply discarded, not so much due to any particular work or polemic—nobody stood up and said, "That's it, folks, crystals are *out*"—but through negligence. Elliptical spheres are kind of hard to nest, after all, and they simply weren't cool anymore.

Still, though, how does it all work? What is the connection between the sun and the planets, between the Earth and the moon, and among the moons of the giant worlds?

The question had been bugging the minds of England's Royal Society, the group partly devoted to serious discourse on scientific matters and partly devoted to drinking, for a few decades. Notable members such as Edmund Halley, Robert Hooke, and others took a ponder or two at the problem. According to Newton, it was his own flash of insight that made the tremendous leap in thought that connected the cosmos together. Of course, we only have his word for it, so make of it what you will.[1]

Outbreaks of plague make it hard for a Royal Society to be a society, and for a university like Cambridge to be a university, so in 1666, cultured life was suspended, and Newton was chilling at his mom's house in Lincolnshire, waiting for people to stop dying so he could get back to work. In the meantime, he walked around thinking all day.

By this time, he had already begun to develop his conceptions of the laws of motion: that it takes a force to make something change its velocity, that the change in velocity is proportional to the force applied and to the object's mass, and that if one object applies a force to another object, then that other object will simultaneously apply an equal force in the opposite direction on the first.

Everyone since there'd been an anyone knew that when you dropped something, it fell to the Earth. But when Newton happened to watch an apple detach from its tree and fall to the ground, he made a connection to his laws of motion—a connection nobody else in the history of anybody had made—and a mental puzzle piece slid into place.

The apple wasn't just falling to the Earth. The apple was *accelerating* toward the Earth. That meant that the Earth was exerting a force on the apple. That force was invisible, but the apple didn't seem to care: it fell. But only in a straight line. It didn't curve or zigzag. This "gravitational force" only connected objects in straight lines, from center of mass to center of mass.

What if the apple fell from a greater height? The force would be slightly weaker, since it would be farther from the Earth. What if the apple were moving sideways when it first started falling? Well, it would still be moving sideways, but it would still fall down.

Now for the big jump—are you ready? What if the apple were as far away as the moon? This "gravity" would be pulling it inward toward the Earth, but if it were fast enough, the apple would stay in orbit forever. What speed would that require?

Presto bingo, Newton was able to follow the logic train to derive the speed of the moon's orbit.

He didn't stop there. Once he realized that gravity might be *universal*, that the same force that pulls an apple from a tree might be the exact same force that keeps the moon in orbit around the Earth, he went nuts. In a good way. Example after example, he was able to show that all sorts of disconnected phenomena across the known universe were really the manifestation of a few simple laws.[2]

What is the source of universal gravitational attraction? Even Newton didn't attempt to go down that road. It works, he argued, so let's just go with it. And the big bow to put on the gravitational present: Newton was able to show that Kepler's laws—the elliptical orbits, the speeds, the harmonies, the whole lot—were a *result* of universal gravitation. One guess about how the universe works was enough to tie together Kepler's entire opus.

Sir Edmund Halley was a big fan of Newton's work, and he went about trying to put the *universal* in Newton's universal gravitation. Halley was also a huge history geek, and if you read any of his astronomy papers, you quickly find yourself being treated to summaries of entire ancient cities. Riveting stuff, if you're into that sort of thing.[3]

The twin passions of astronomical minutiae and historical minutiae led Halley to some seriously non-minute conclusions. You may already be familiar with him from his famous comet, whose reoccurrence he predicted by noting a pattern in the historical record and using universal gravitation to tie it together.

He also totally nailed the prediction of an eclipse to hit England in 1715, which gave him instant celebrity status around the country.[4] Solar eclipses were notoriously hard to predict (as the ancient Chinese astronomers found, to their headless dismay) and the attempts of our ancestors to forecast them based on complicated and interweaving patterns, subpatterns, almost-repeating cycles, and exceptions to the rules is almost sad. They tried so hard, but they couldn't quite crack it because they simply didn't have the right tool.

With universal gravitation, though, Halley was able to predict the next total solar eclipse to within four minutes. In the eighteenth century, that's practically atomic-clock-level accuracy. He even made handy-dandy maps detailing what you would see when and where. If you've paid any attention at all to modern-day maps of eclipse paths, you can thank Halley for setting the standard. He nailed that sucker.

Just as easily as Halley could turn his newfound superpowers to predicting the future, he could use them to understand the past (remember, he was a history dork). He was especially fascinated by records of eclipses and liked making maps of what ancient peoples would have experienced during those events.

The oldest one he could get his nerdy little hands on stretched back to about 900 BCE in the Middle East, after he interpreted (and corrected!) the translations and retranslations passed down through the centuries. And he spotted a slight, niggling issue.

Flexing his universal gravitational muscles, Halley could handily run the clock backward and compare predicted (postdicted?) eclipses to the actual historical records. At first everything was bang on, with each result of Newton's laws matching what folks wrote down so long ago. But far enough back, errors started to creep in, and the further he pushed into the past, the greater the divide between theory and experiment.

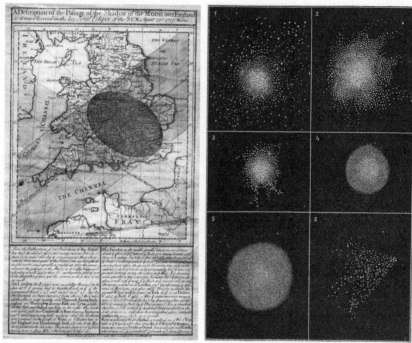

Understanding and confusion II: *Left*, Halley's amazing achievement in accurately predicting the 1715 total solar eclipse and giving an eager public a detailed map of the event. *Right*, more than 150 years later, Sir Norman Lockyer's sketches of various nebulae and clusters still defy explanation.

Halley didn't really know what to say about it. Newton's universal gravity was so gosh-darn universal that it was hard to discount it. But the historical record was the historical record. Assuming there wasn't some giant millennia-spanning conspiracy to fudge the eclipse records, he had to take them at face value.

Halley added a brief note as a closing remark to a long treatise on the long-dead city of Palmyra (you can try to visit the ruins in modern-day Syria), along the lines of "Hey, guys, I think the moon is doing something funny, but I haven't confirmed it yet, so hold on. Be right back."[5]

And he never brought it up again.

But others did, and they confirmed Halley's suspicions: by carefully combining Newton's laws with the historical record, they could deduce that eclipses were slowly getting further apart.

After a bit of math (well, truthfully, a metric ton of math over the course of a few decades, not getting fully resolved until the mid-1800s), the answer was worked out. Indeed, the moon was slowly receding from the Earth, prolonging the duration between eclipses. That recession is caused by the same tides that the moon is responsible for.

When the moon is overhead, a lump of water rises up and tries to meet it: a tide is born. But the Earth is spinning, so it carries the tidal lumpy bit farther ahead of the moon's position. That leaves a giant blob of mass sitting "in front" of the moon from its perspective, and that lump, being massive, pulls on the moon, as gravity is wont to do. Like an invisible gravitational leash, the tide tugs on the moon, giving it energy and booting it to a higher orbit.

That means in a few hundred million years, the moon will be small enough in our sky that total solar eclipses will be impossible. So enjoy them while they last!

This is fine and dandy. Indeed, it was another spectacular result for Newton's brainchild. But what it meant about the *past* was a little more troubling. If the moon is moving farther away from the Earth, then simple kindergarten logic dictates that it used to be closer. And in the distant past it was so close it must have . . . touched . . . the Earth?

The universe was different in the past. And not just a little bit—wildly, fantastically different. So different it defies logic and common sense. It's a big pill to swallow, and that was a big reason for the objection to even *working* on the eclipse problem for a few decades. But eventually the math won out, as it usually does, and everybody had to accept that fact.

Their only solace was that you have to go waaaaay into the past for the moon to be anywhere near Public Displays of Affection distances to the Earth, like hundreds of millions of years into the past. And there's no way the Earth could be that old, right?

Right?

By 1800 William Herschel was already a superstar. There are only three people in all of human history who can lay claim to discovering a new planet in the solar system, and one of them (Clyde Tombaugh, who discovered Pluto in 1930) was later disqualified on a technicality. In 1781, Herschel was the first to grab that title,[6] and had it been me, in all honesty the seventh planet of our home system would be called Sutter's Awesome Planet. But Herschel wasn't me, so after a few rounds of suggestions everyone settled on Uranus, the Greek god of the sky, thereby ensuring that generations of English-speaking school kids would have something to giggle about when memorizing the planets.

Just let that soak in for a moment. No, not the Uranus puns—the concept of a *new planet*. Planets are pretty easy to spot, if you're dedicated enough. They are the "wanderers"; they move, ever so subtly, across the background of the distant stars from night to night. Uranus itself is faintly visible to the naked eye on a clear, dark night (which the ancients had in abundance), but unless you're *really* looking for it, it's easy to miss.

Herschel wasn't exactly looking for it—he was hunting for ever-fainter stars—but he did notice a discrepancy between different observations. And almost overnight, our cozy little planetary family added a new member. I don't know how pre-Copernicus thinkers would have handled the discovery of a new planet. Just added another crystal sphere to ferry the new celestial denizen? Updated all the astrological charts with signs and portents and significance? "Oh, *that's* why we didn't predict you would get smallpox—we were missing the influence of *Uranus!*"

We'll never know, because Uranus was discovered in 1781 and not 1581, and everybody went crazy with the news ("news" was also now a thing) and Herschel was an instant astronomy legend.

Nineteen years later, he was playing around with light. A couple of generations earlier, Newton had already shown that white light was really a mixture of all the colors. A simple prism is enough to demonstrate the effect, but what Newton showed was that a prism wasn't *creating* the colors from white light but simply *separating* the colors already inherent in the beam.

Herschel got the bright (ha!) idea to measure the temperature of bits of light: Is red hotter than blue? Or vice versa? Or the same? Good old-fashioned science-type questions that only a science-type person would be bothered to (a) ask and (b) actually try to answer.

So he split a beam of sunlight using a prism and started sticking homemade thermometers on various colors and dutifully recording the results. Ever the careful observer, he put thermometers on either end of the rainbow as an experimental control.

But control it did not. Herschel noticed something funky going on: the thermometer sitting outside the red part of the spectrum was warmer than any other color! And it wasn't just a freak accident of experimental design: he started playing with these invisible "colorific rays" (a fancy term for "heat rays") and discovered they did all the same stuff that normal light did. He could reflect them, refract them, absorb them with certain materials, transmit them through others, and on and on. These rays had all the same properties as light; they were just redder than the reddest thing we could possibly see with our eyes. *Infra*-red, if you will.

With a one-two punch, Herschel knocked our knowledge of the universe on its back: A new planet! And a new kind of light!

The cosmos was getting complicated, fast.

The telescope wasn't helping the situation at all, but at least folks like Charles Messier were taking the time to write things down. Take a moment to think about the sky that you see in your backyard with the naked eye versus what sky even a small telescope reveals. Galileo almost had his mind blown by his crude instrument's portrait of the heavens—shapes, textures, and depth that our lowly iris simply can't capture.

The fixed stars (though as we quickly learned, they're hardly "fixed") weren't stuck to the outermost celestial sphere. Pick an empty patch of sky. It looks like nothing's there: pure, velvety, smooth blackness. Point a telescope there. What do you see? Stars. Loads of them. Pick an empty patch among *them*. Get an even bigger telescope and point it there. What do you see? No points awarded for guessing the correct answer.

The number and variety of creatures inhabiting our universe grew with every decade. A menagerie of comets, nebulae, multiple stars, other kinds of nebulae—it went on and on. It seemed endless and bountiful and utterly confusing.

Take just the nebulae, for example. Taken from the Latin word for "mist," the name stuck for obvious reasons. If you see something in the sky that seems to be (a) far away and (b) not a star, it's a nebula. Some you can see with your eye, but most can only be viewed with an astronomical helper. And it's a sampler box out there: all manner of shapes and sizes and a dazzling array of colors.

Just check out the Messier catalog, a list of fuzzy objects that definitely aren't comets compiled by French astronomer Charles Messier in the later 1700s.[7] Comet hunting was big business in those days, and so many excited astronomers were ecstatic to find something new in the sky but quickly disheartened to learn it was not a new comet but an already-identified fuzzy patch.

Messier wanted to fix that (probably mostly for himself, as he was a comet spotter extraordinaire, but it also turned out to be useful for other people), so he listed, in no particular order, a collection of fuzzy things. Some were really just clumps of stars. Some were mostly round and bland. Some had strange helical patterns and interwoven colors. Some were vast, with vague spiral-like appendages. They were all beautiful—there was no doubt about that—but they were downright mysterious.

This theme—"Let's explore the heavens with no clue what we're looking at"—resonates throughout the nineteenth century. The instruments of astronomy had advanced way beyond the capabilities of astronomers to understand their own observations. Problems mounted and intensified. Ever get hungry but not know what you're hungry for, and your indecision only makes the hunger grow? The 1800s were like that, but for science.

Here's another example: the rings of Saturn. First spotted by Galileo himself with his homespun optics, they appeared as two lumps on either side of the great planet. Over time they would flatten and disappear, only to return later and fatten up again. The very next generation of astronomers after the Italian realized that they were looking at a disk and Galileo's frustrated observations were caused by alignment: sometimes he would see them face-on, and other times edge-on, depending on our position relative to Saturn in the solar system.

But to him, it was just question marks all the way across the page. "Has Saturn swallowed his children!?" he wrote in a letter, perhaps only half-jokingly referring to the Greek myth.[8] By the late 1800s the mystery was still unresolved (astronomy joke, sorry). We knew that it was rings, and that there were gaps, and that it wasn't a solid disk but made up of smaller particles. That last bit was proven by the great James Clerk Maxwell, the genius who united the forces of electricity and magnetism into a single unified description—electromagnetism—and also basically discovered light. Smart dude, right? As to his explanation for the cause and composition of the rings? Got nothin'.[9]

For those of you keeping track: no, we still don't fully understand the rings of Saturn today, despite having Hubbles and spacecraft.

I'm confident that Joseph von Fraunhofer wasn't planning on completely revolutionizing the field of astronomy when he got too caught up in staring at the sun in the early 1800s, but he totally managed to do that, so here we are.

We already talked about how Newton demonstrated that white sunlight was really a mixture of all the colors of a rainbow, which leads to a very natural question: how in blazes does the sun produce all the colors of the rainbow? If you hold a candle up to a prism, you also get a rainbow effect. So now you know that the sun, like a candle, is both hot and glowy. A somewhat mild accomplishment, but an accomplishment nonetheless.

Working in relative ignorance as the *why* of rainbow, Fraunhofer (and others before him) decided to tackle the *what* in more detail. By passing the prismed sunlight though even more prisms, he could spread the light out farther than anyone had before, and it was in the enhanced sunlight *spectrum* (because the word "rainbow" doesn't sound sciencey enough, I guess) that Fraunhofer found something fishy.

Specifically, he saw something missing. Embedded in the spectrum of sunlight were hundreds of distinct dark lines, no wider than a hair, at seemingly random places within the colors. So our sun isn't giving us, for example, 100 percent of the color yellow—we're only getting 99.9 percent of that color, with very specific wavelengths nibbled out.

But these wavelengths are not missing from the spectrum of a simple candle flame. Aha: the sun isn't quite what it seems to be.

It wasn't for a few more decades that a puzzle piece clicked into place when Robert Bunsen (of "the burner" fame) and Gustav Kirchhoff (of "who?" fame) figured out that when specific elements were tossed into a flame, *bright* lines would pop out of the flame's spectrum. It's as if every element has a fingerprint—a pattern of lines in an otherwise featureless spectrum that is unique to that element.

Perhaps—work with me here, guys—when an element adds its light, we get bright lines appearing, but when an element blocks a background light, the same lines appear, but dark. Like a cosmic crime scene, the spectral fingerprints point the way to the suspect elements.

And now we can figure out what the sun is made of. And the stars. And nebulae. And planet atmospheres. And . . . everything. Granted, at this time nobody understood *why* the elements produced these strange lines (and even the concept of "element" was still gaining ground in chemistry circles), but what mattered was that they *did*, and we could test that in a laboratory in the back of the office.

That's kind of a big deal. This technique, known as spectroscopy (because, again, "rainbowscopy" doesn't sound sciencey enough), is the ultimate key that unlocks the farthest reaches of the universe. We can taste the surface of the sun without having to visit it, simply by comparing sunlight to various heated gases. We can discover brand-new elements, as Jules Janssen and Joseph Lockyer did in 1868 when they identified in sunlight a never-before-understood series of lines as helium, which they named in honor of our own sun.

It wasn't all roses and unicorns, though. Every new discovery in science leads to a thousand more questions. Hey, there's oxygen in that nebula way over there, awesome! Wait—how did oxygen get into that nebula? Way over there?

It gets even better/worse. Once folks started to come to terms with the concept of light as waves (waves of *what* would have to wait until 1865, when Maxwell realized that they're waves of electricity and magnetism), there was another trick that spectroscopy could play. If you've ever heard something loud passing by, you're familiar with the Doppler effect: on the approach, sound waves get squished, pushing them into higher pitches. On the way out, sound waves get stretched, pulling them to lower tones.

It happens with sound waves, and it happens with light waves. It's a very subtle effect, though, since most things don't move very fast compared to the speed of light itself, so it's not like we see entire colors shifting redder or bluer. But the fingerprint pattern of spectral lines can shift. It's a wonderful tool: perhaps you recognize a particular arrangement in the spectrum of a star— oh, there's some iron!—but the whole pattern is shifted to the left or right by a few wavelengths. Well, if the spectrum from that star is shifted toward the blue end compared to something stationary, like a light source in the room you're sitting in, not only can you conclude that the star is moving, but you can measure very precisely its speed.

Not its entire speed—sideways movements won't change the spectrum from our perspective—but the in-out speed is fair game for measurement. And measuring the speed of star after star revealed that we don't live in a fixed cosmos. We live in a beehive.

There's one other piece of technology that opened a window of confusion to our universe: photography. Where the telescope acts as a super-eye that creates a bigger bucket to collect light and magnifies separations to make them more distinct, adding a photographic plate to the back end of that device amps it up to a hundred. No matter how good your telescope is, if you only look with your eyes, you'll be fundamentally limited by what you can see.

But a photographic plate can collect, collect, collect. Restless and unblinking, it continually absorbs light, adding it to the pile, revealing fainter and dimmer objects. And it records! No longer do you have to alternate between staring and sketching to record what you're seeing—the photograph does it all in one handy-dandy, convenient device.

Astrophotography is the ultimate extension of the human sense of sight into the cosmos. It's everything a human eye does, just way better. Combined with spectroscopy—the study of spectra—it opened up the heavens like never before. Information poured in from observatories across the globe. Expeditions were launched; telescopes were fashioned by professionals and enthusiasts alike. Never before had so much interest been focused onto the night sky. The number and variety of phenomena in the universe around us were almost overwhelming. Breathlessly, astronomers recorded and published their findings in between sessions of staring dumbstruck at their celestial revelations.

Over the course of the eighteenth and nineteenth centuries, we discovered and cataloged new kinds of nebulae. We confirmed that the distant stars were like our sun but also different. Some were smaller, some larger. Some hotter, some cooler. Comets came and went on repeatable cycles. Our solar system was belted with a ring of asteroidal debris. Dozens of moons danced around the outer worlds of our solar system. Dust glinted in the pale sunlight and was flung out between the stars themselves. The universe itself was beginning to open up before us.

I'm always hesitant to pull random quotes out of history just to mock them, because it's kind of challenging to predict the future, but this one is too juicy to pass up. In 1835, the philosopher Auguste Comte wrote, "I regard any notion concerning the true mean temperature of the various stars as forever denied to us" due to their extreme distance.[10] He wrote many other things that turned out to be useful and respected, but in this one instance, the scientific community, after decades of labor, analysis, and careful study, responded with a resounding "Bite me."

But nature has a habit of biting back, and I'm sure Heinrich Olbers thought he was making a really great point in 1823. Turns out he was far from the first person to have this thought, but he was pretty much the last, so his name got stuck to an apparent paradox.[11] The paradox goes back to the old sun-centered versus Earth-centered arguments of generations past. In the solipsistic view, where we here on Earth are the literal center of the universe, the fixed stars are simply that: fixed. There are a few thousand of them, and they're all attached to the outermost celestial sphere, wheeling away through cosmic time. No big deal.

But in a sun-centered universe, you have to grapple with the distinct possibility that the fixed stars are really distant suns. And as soon you point your telescope into the deep void between them, you find other, fainter—and possibly more distant—fireballs.

So how far back in space does it go? How deep could we possibly perceive? What lies in between the in-between? Do stars stop? Is there a limit to their light? If that's the case, does that kind of universe even make sense? To have a cosmos filled with void, except a little cluster of lights in one corner?

Isaac Newton provided one answer. If the universe were finite in space, then eventually gravitational interactions would, over the course of uncountable eons, cause all the stars to accumulate into a single pitiful lump.

The easier thought to swallow is that the universe is infinite. It simply goes and goes, with countless stars backed by ever-larger multitudes. Thus any spot in space is perfectly gravitationally balanced by the infinity on either side.

But how far back in time does it go? Various religious traditions teach about the (re-)creation of the physical universe at distinct points in the past, but if there's one thing the Copernican revolution taught us, it's that maybe we should give the scientific method a chance at answering some of those questions.

The Earth isn't going anywhere soon, and neither is the sun. And neither are the stars, or comets, or nebulae. They're just *there*. Sure, a new star will occasionally appear, or a comet will break apart upon encountering the inner solar system, or the moon might be slowly circling away from us, but for the most part, the universe today looks like the universe yesterday. And the day before, and the day before. Maybe the universe is simply infinite both in space and time. Maybe there is no beginning, no primordial ooze, no "Let there be." The universe *is*.

But that doesn't work, and that's where Olbers' paradox comes in. In a universe with infinite extent in both time and space, there shouldn't be any dark. If you look in any random direction in the sky in an infinite universe, then you *must* be looking in the direction of a star, somewhere, at some distance. "But maybe the light hasn't reached me yet," you retort. Good point, except that in a universe that has existed for eternity, there's been way more than enough time for that light to reach out.

So night shouldn't be night at all; instead it should be aglow with the fire of literally an infinite number of suns. But it's pretty dang dark, which means the universe isn't infinite in time and/or space. But all the lines of thinking and evidence point toward infinity. What gives?

It will take me a lot of words to fully deconstruct the apparent paradox in detail—and don't worry, I certainly will in later chapters—but the short version is that the universe is definitely not infinite in time (at least, into the past) and most likely not infinite in space. But our dear nineteenth-century friends didn't know that, so they had to grapple with the central conundrum.

They attempted to tackle it one step at a time, and the first step is getting a distance to a star: any star at all will do. Just give us one hook into the extra-solar system, and we can start putting together a map of the cosmos and figure out the flaw in Olbers's reasoning.

I'm never one to call Newton naïve, but he did advocate a naïve method for measuring distances. If you assume that all the stars are the same *true* brightness—in essence, that they're all identical copies of the sun—then if you can measure their *perceived* brightness with incredible precision, you can do some math and figure out a distance. Astronomers over the following decades confirmed that most stars are totally unlike our sun in color and temperature and elemental composition, so that's a nonstarter. The method isn't totally without merit, and later generations will use the same principles to great effect, but that's a story for a later chapter.

Instead they had to give parallax a try. Parallax is the simple geometric measurement where you pick a star, record its absolute position in the sky, then wait six months until the Earth is on the opposite side of the solar system. Repeat your measurement, and if you're very good and even luckier, you'll have recorded a small shift in its position. That gives you an angle, and since you (hopefully) know the distance to the sun, you can construct a long skinny triangle toward the star, do some basic trigonometry, and compute a distance.

Simple, but not easy. We first encountered Tycho Brahe himself attempting a parallax measurement to put the nail in the coffin of all this sun-centered universe nonsense. He succeeded: according to the very best measurements the world had ever produced (ahem, his own), there was no observed parallax, and hence if the sun were the center of the cosmos with the Earth flinging itself about it, that would mean the celestial sphere had to be . . . let's see here . . . seven hundred times farther away than the planet Saturn. Preposterous that the universe should be so large! Earth-centered it is, chaps.[12]

Even though Kepler and then Newton won the sun-centered day based on other arguments, the problem stuck. Surely somebody would eventually measure a reliable parallax, get a fix on a star, and start to put the *nomy* in astronomy. But decades, and even centuries, churned by without a measurement. Telescopes got bigger and better. Catalogs of the heavens grew thick with entries. Innovative techniques were developed and deployed. The heavyweights I've already introduced in this chapter all took a crack at it.

Nothing. Not a single distance. With every failed attempt, we had to stretch the yardstick of space out farther. With every false report, the universe grew larger: the greater the distance to the stars, the smaller the seasonal wobble, and the better our instrumentation had to be. It was getting kind of scary, honestly.

Finally, after centuries of previous attempts and years of his own hard labor, Friedrich Bessel nailed it: 61 Cygni, a star in the constellation Cygnus that's unremarkable except that it's close to Earth. Bessel didn't *know* it was close, but he guessed based on its larger proper motion over the decades and centuries. ("Proper" here is a bit of astronomical jargon to mean motion that belongs to the star itself, not due to any "fake" motion that we might observe from the rotating vantage point on the surface of the Earth.)

It was already realized that stars move of their own accord, even before later spectroscopic measurements would confirm it. Since the stars are very far away (as has been established), it takes time for their motion to be noticed, but noticed and measured they can be, and it was (correctly) argued that if a star is closer to the Earth, it should have a bigger proper motion, because that's how geometry works: cars moving across the intersection right in front of you will have greater proper motion than ones a few blocks away, even if the cars are all moving at the same speeds.

61 Cygni is one of the fastest stars, so Bessel figured it would give him a shot of winning the big prize, and he was right. In 1838, after a few years of observations, he came up with a distance of ninety-six trillion kilometers. That's right, "trillion" with a terrible *t*.

Remember, just a couple of centuries earlier, Tycho Brahe was nauseated by the thought of the stars sitting seven hundred times farther from the sun than Saturn. Bessel's measurement, which is only 10 percent off from the current best measurement, placed 61 Cygni about sixty thousand times farther than Saturn.

In one clean measurement, Bessel (who, I feel compelled to note, received no higher formal education and also managed to develop suites of mathematical functions that bear his name today) finally put to rest the ultimate question of the heliocentric debate. It was already on firm theoretical grounds thanks to Kepler, Newton, and others, but this was a key piece of data that had been missing from the arguments.

Other parallax measurements quickly followed (Bessel wasn't the only one interested in the problem). The universe was getting larger and more complex with every new telescope and every new catalog. It was a heyday for the experimentalist and a nightmare for the theorist. None of it made any sense: just how big is our home, and what is it made of? These stupidly simple questions were getting frustratingly hard to answer.

At about this same time, astronomy was finally splitting off from astrology. For millennia, the two had been intertwined, and the words were roughly synonyms. Measurements of the stars went hand in hand with their forecast effects on our daily lives. In a complex, chaotic world where nothing made sense and everything was changing all the time, it was no wonder our ancestors looked to the steady, regular patterns overhead and drew solace from them. The clockwork regularity of the stars and planets must hold the keys to the underlying order of life on Earth.

But by the close of the nineteenth century, we knew that the universe at large was as frightful and complex as anything here on Earth—and even more so. There were forces and motions at play too great to comprehend. 61 Cygni itself was computed to have a proper motion of hundreds of thousands of miles per hour. How does a star, a massive burning ball of gas, achieve such incredible feats? How can new stars appear and familiar ones go silent? How can Newton's laws account for all this?

The telescope, the spectrometer, and the photograph opened up the cosmos before us, but it was a cosmic Pandora's box. We struggled and grasped to connect the physics we were learning on the Earth—electricity and magnetism, heat and energy, chemistry and the element, and other hot topics of the day—to the scales of the heavens, and we failed, terribly.

It was becoming painfully obvious that the cosmos was not connected to us, did not care about us, and did not care for us. We were an ant climbing on a branch of a vast tree that was incomprehensibly larger and more complex than we ever thought. We were reaching out with our enhanced senses and the powerful tool of the scientific method, and we were not liking what we were seeing.

While astrologers clung—as some still do today—to the notion that the motions of the planets govern and predict our lives, astronomers were left in a much more befuddled state. They could record, measure, and study, but they could not understand.

By the early twentieth century, astronomers were especially concerned by the nature of the spiral nebulae—just one branch of that fuzzy family tree, but one that seemed to be different from the others. How far away were they? Were they part of our universe—an unbroken field of stars stretching from one end of the cosmos to the other—or somehow isolated from us in their own "island universe"? Data and argument swung either way depending on who was more persuasive and whose data you believed to be more reliable.

Our understanding of the universe was at a breaking point. A hurricane of raw data was slamming into previously held notions. We couldn't crack the code; we couldn't navigate the storm of conflicting ideas and theories.

Our perception of the cosmos was ready, if you will, for a phase transition.

CHAPTER 4

THE DEATH OF ANTIMATTER

I want you to imagine visiting a hotel, one familiar but with some odd properties that you usually don't encounter. Feel free to call it *Hotel Dirac* if you want. If you don't, you'll get the joke later.

Here's how the hotel works. There are multiple floors and rooms on each floor, as usual. But there can only be one person assigned to any room at any one time, and the rooms are filled starting with the first room on the first floor, then the second room, and so on, until the lowest floor is filled; then rooms get assigned on the second floor. There's technically an infinite number of floors, but that's not really relevant.

Sometimes a guest will feel like moving up in the world and bump themself up to a room on a higher floor. But they only get to visit that penthouse suite for a little bit of time—as soon as one of the hotel staff notices (and they're very diligent about these sorts of things), the guest gets scolded and pushed back down to a lower room.

Now here's where the hotel gets especially strange. The rooms don't stop at the ground floor—there's a basement, and that basement is full of rooms. And beneath that is another subsurface floor, equally stuffed full of rooms. And down and down and down—there is technically an infinite number of rooms *beneath* the surface, but that's not especially important for now either.

When you arrive at Hotel Dirac, the hotel may look empty, but really all the subsurface floors are already occupied with guests, one per room. So what you see as an unoccupied ground floor actually sits on top of countless millions of underground guests.

When I say that any guest in the hotel occasionally moves up to a random higher level, I really mean it—that includes the underground rooms. Someone down there can get motivated, find the nearest elevator, and briefly get to enjoy the views out the window—until the surly hotel staff finds them and bounces them back down.

When such a guest gets bumped up to a higher room, they may find that all the room doors are unlocked—they can wander from room to room on their floor, flicking on the lights as they go, checking out the layout and seeing all the possible views. They can't change floors without permission, but the rooms on each floor are fair game.

What about the empty room they left behind, on one of the underground floors? Down there all the rooms are unlocked too, and curious looky-loos will float in and out of that unoccupied room, one at a time.

What does Hotel Dirac look like from a distance when an underground guest gets bumped to a higher floor? That guest, floating from room to room, turns on the lights when they enter and off when they leave. From far enough away, you'd see a single light on an upper floor, shifting from room to room as the guest explores.

If you could see through the ground, what would the subterranean floors look like? As the guest left, they dutifully turned off the light—that room looks like a hole in an otherwise unbroken sea of lights. But as other nosy guests switched to the unoccupied room, the "hole" would appear to shift around.

For as long as the guest got to remain upstairs, a light would move on the aboveground floors, and a hole would move on the underground floors. Once the staff noticed the discrepancy, however, the guest would reluctantly slink back down beneath the ground level, find the nearest unoccupied room, and turn on the light, closing the hole and returning life to normal.

This is the picture of the world as painted in the 1920s by Paul Adrien Maurice Dirac, who accidentally discovered antimatter as a result of trying to solve another problem—reconciling the burgeoning description of quantum physics with the already-established theory of special relativity.[1]

This marriage was attempted and then abandoned by Erwin Schrödinger, who instead settled for a less general formalism—his eponymous wave equation, which students across the world grapple with in frustration on a daily basis.[2] But Dirac managed to nail it; all it took was a completely new mathematical description of the world and a serious rethink about the nature of reality, so you can see why Schrödinger shied away.

And buried within the mathematics of the theory was a surprising little symmetry: a new *kind* of particle, positioned like a mirror to our everyday world. Every fundamental particle, like an electron, was matched by a new particle with identical properties (mass, spin, etc.) but with perfectly opposite charge. For the electron, its *antimatter* twin is called the positron. Exactly like the electron, but with a positive sign in front of its charge.

The positron was experimentally discovered a couple of years later, followed shortly by the twins (antitwins?) of all the other known particles.

The scenario of Hotel Dirac is a useful way to explain a surprising situation: if you have a bit of light—a photon—at high enough energies, it can spontaneously split into an electron and a position. After traveling for a bit, the two particles will find each other again, collide, and disappear in a flash of light, releasing back the original photon.

In this picture, what we view as the "ground state" of the electron in any situation—the lowest possible energy state—actually sits on top of an infinite pancake stack of negative energy states, already occupied by a subterranean hotel full of electrons. A photon of sufficient energy can knock one of these negative-energy electrons into a positive-energy state, where it runs around doing all the things that electrons do. But it leaves behind a "hole" in the sea of negative-energy electrons, and that hole looks, acts, and smells like a typical electron—except it has the opposite charge.

Eventually the electron gets tired of wandering in the positive-energy world and falls back down into its hole, releasing the energy that originally promoted it. The photon returns, and everything is back to status quo.

This picture isn't exactly correct—our more modern view of the process is different in some subtle and important ways, and I'll get to that in a another chapter, but it does serve a very useful point here: matter and antimatter are symmetric. Or at least, *ought* to be symmetric. For every piece of matter—an electron over here, an atom over there—there ought to be a matching twin with the opposite charge out there, somewhere.

This symmetry of antimatter is baked into the same mathematics that predicted its existence in the first place. It appears completely unavoidable.

But look out there, somewhere, anywhere. See any antimatter? No, you don't. From one end of the Milky Way to the other, from the earliest moments of the universe that we can observe to the present day, matter rules the cosmos. Almost all the *stuff* is normal, not anti. If there were, say, a galaxy composed entirely of antimatter, then as it swam through the thin soup of particles between the galaxies, it would be releasing enormous amounts of energy—the most energetic events ever known.

One ounce of antimatter annihilating with one ounce of normal matter would release the energy equivalent of about a good-sized H-bomb. One galaxy of antimatter annihilating with one galaxy of matter would release—let me see here—ah, right, a *lot* of energy.

We don't see it. We don't have the suspicion of seeing it. We don't even have a *hint* of a suspicion of seeing it.

Matter, not antimatter, dominates the universe, and has for an incredibly long time—essentially its entire history. We're obviously misunderstanding something.

Where did all the antimatter go?

When we last left the story of the universe, it had reinvigorated itself after exhausting its energy in the most rapid expansion yet known—and ever to be known. Inflation had ballooned the observable cosmos to the preposterous dimensions of an apple, in the slightest sliver of a second. That inflation was triggered by the splitting of the strong nuclear force away from the others, and the process spread out all the matter into a cold, thin soup.

This soup was *somehow* reinvigorated as whatever drove inflation shook itself off (cosmologists are still working out that detail of the story). But that reinvigoration itself creates a new problem (sensing a common theme here?). Let's say the universe at this stage is filled with high-energy radiation at a temperature of 10^{15} K. Some of those photons can transform into pairs of electrons and positrons (and a host of other particles), but they will always do so symmetrically. For every bit of matter that pops into existence, a matching bit of antimatter will be along for the ride. Eventually they'll find each other, tragically end their lives in a fury of mutual self-destruction, and return back to radiation.

But our universe, today, is not filled with only radiation. There's matter all over the place. To create the amount of matter that we observe, the imbalance in the newborn universe didn't have to be much: just one part in a billion extra in the ratio of regular matter to antimatter would do it.[3] That's a small number, but even small numbers are huge compared to the *totally zero* predicted by particle physics.

Long side note: Unfortunately, when dealing with high-energy physics, which is the realm of the early universe, the jargon comes fast and furious. We have to keep track of the names for all the particles, the antiparticles, the hypothetical particles, the forces, the hypothetical forces, groups of particles, families of particles, and on and on. Plus it's all twisted up because some processes got their names assigned before we fully settled on a definition. Hence I've been trying to avoid most of that messiness, but just in case you want to look this stuff up on the internet later (you masochist), the name given to the process of the domination of matter over antimatter is *baryogenesis*, and no, my spell-checker doesn't recognize that as a real word either. *Baryon* in most contexts means particles made of three quarks, like the familiar proton and neutron. And quarks are—well, I'll just save that for later.

Anyway, to solve this riddle, we have a few options. Take your pick:

Option 0: Asymmetry is a lie! There is actually tons of antimatter out there; we just live in a little patch of regular matter. But as I talked about earlier, the implications of that kind of universe seem kind of (a) violent and (b) obvious, so essentially nobody finds this palatable.

Option 1: The universe had an asymmetry between regular matter and antimatter baked in since the beginning. I know, I know, there may not be a "beginning," but in this argument, there is some overarching rule that says the two kinds of matter are not really in balance, and it's been in place for the entire history of the universe, in a similar vein as a rule that says, "By the way, there's a force of gravity." This is generally unappealing because it's totally just sweeping the problem under the rug and pretending it doesn't exist, and because we see no evidence of this grand law operating in the present-day universe. You would think something that was that big of a deal ought to hang around for longer than a second.

Option 2: Hey, I know, there were lots of crazy physics happening in the preinflation madhouse era, so maybe that's the key! We don't really understand the physics, so maybe tucked into an equation here or slipped into a term over there is an imbalance, and that will do the trick. In this story, before inflation even got rolling, the stage was set for baryons to dominate. Our understanding of this epoch is fuzzy enough to accommodate lots of wacky ideas (something the theorists among us love) but not clear enough to actually separate one idea from another (something the theorists hate, because none of them can get the validation to win a Nobel Prize). While a valid choice, this option is basically the community saying, "Let's have the *next* generation of scientists solve this one."[4]

Option 3: Maybe the imbalance came later, after inflation, as the universe was steadily expanding and cooling. It's still a crazy mess of a place, and there's plenty of particle wiggle room to get up to some funky stuff. For example, the weak nuclear force still hasn't split off from the electromagnetic force, and while we largely understand that process, there might be something interesting there.

Just for fun, and because it's our most solid lead, let's follow Option 3.

I really need to introduce the weak force properly to show how it might play a rather unexpected role in disrupting the delicate balance between matter and antimatter. Let's face it: nobody treats the weak force with any respect. I mean, just look at the name! The other forces have ancient, complex, or assuredly self-descriptive names. The weak nuclear force is indeed weak, but it's far, far stronger than gravity. And it does play a role in nuclear reactions, but not in the same way as the strong force.

In essence, the strong nuclear force is a *binder*: it glues things together (except when it repels—it's complicated[5]). The weak nuclear force is a *transformer*: it can change one kind of particle into another. That may not seem impressive, but it lies at the heart of radioactive decay and the synthesis of heavy elements. So yeah, kind of important.

And when it comes to matter versus antimatter, the weak nuclear force has a favorite. It's not immediately obvious, and for the effect to show up it requires piles of particle collision data. It also came as a big surprise to particle physicists when it was discovered in the 1950s and '60s, but that's just life.

I won't go into the details here, since I'm trying my best not to make this a textbook on particle interactions,[6] but in a collider you can make some exotic combinations of particles. These exotic combinations don't hang around for long—they're unstable and quickly decay into a shower of smaller, longer-lasting particles. Two of these bizarre characters, called the pion and kaon (pro tip for any wannabe particle physicists: if you need to name something new, just take letters from another alphabet and add "on"), decay into various children particles with various rates.

But they don't decay into exactly the same particles at the same rates every time. They show an ever-so-slight preference for decaying into one combination of charges versus the opposite. In the jargon that we are now enmeshed in, their decays violate *C-symmetry*, or symmetry of charge. These decays also violate another apparent symmetry of our universe called parity, which means that all fundamental interactions at the deep particle level look the same in a mirror. Well, almost all: these pion and kaon decays are an exception.

So both C (charge) and P (parity) are not essential symmetries in the cosmos, even though for a long time we thought they were. If you're curious, the ultimate combo of charge-parity-time (CPT) *is* thought to be persistent: if you take a particle interaction of your choice, flip all the charges, run it in a mirror, *and* run it backward in time, you shouldn't see any difference. But let's not get carried away here.

This symmetry violation in the weak nuclear force is important because it provides a *known channel* for favoring one kind of electric charge. And since the weak nuclear force doesn't get to become a player in cosmic history until *after* inflation, that's why we think it's a prime candidate for making the universe— wait for it—matter.

Violating this central symmetry, however, isn't the only part to the story. I know, just as you thought we were getting out of the woods. Two other conditions must be satisfied if you want more matter than antimatter, and they are harsh conditions indeed.

One is that there must be a process that produces a raw excess of matter over antimatter. Wait, what? Why didn't we just, you know, start with that? You might think this is the only condition you need, and you're almost certainly questioning my decision to regale you with tales of kaons and symmetry violations. However, you need *both* conditions (a favor for matter and a favor for charge) to get the desired result.

Now would be a good time to take a break. Go on, I'll wait.

You can have a process that makes an abundance of matter all you want, but if charge symmetry is enforced, it *must* be matched by a process that generates more antimatter than the regular kind. In other words, you might think you're cleverly generating an imbalance, but nature will sneakily slip in some back-channel reaction when you're not looking to make sure everything evens out. Then all the particles will end up back in balance, despite your best efforts, and you'll be stuck with a radiation-only universe again. So you need a channel for generating extra matter, and you need to make sure that channel isn't negated by its evil charge twin. Only then can you flood the universe with regular matter.

Unfortunately, creating excess matter doesn't happen in our normal everyday universe. Fortunately, the initial moments into the history of the cosmos aren't our normal everyday universe.

The culprit is once again the crafty weak nuclear force. We know that every once in a while, a weak interaction can produce an excess of matter, but the channels available to do it are highly suppressed—they are so rare that they essentially never happen at low energies. But at high energies, especially energies high enough to merge the weak and electromagnetic forces, these processes can operate at full blast. So it's certainly possible to transmute radiation into a matter-filled early universe, using hidden tricks and trapdoors built into the nature of the weak force itself.[7]

Unless your universe is in equilibrium. If you have a hot ball of gas or plasma, and it's left totally to its own devices, then all allowable processes and interactions within that hot ball will happen, canceling out each other. That's the very definition of *equilibrium*. So if the young universe were in such a state, any method that produced extra normal matter would be competing against other methods that produced extra antimatter, and nobody would win, ending in a draw (i.e., no matter at all, anywhere).

Back to square one.

Ever ready for a three-peat, the humble weak force comes to the rescue to satisfy this last of the necessary conditions to tip the cosmic scales in favor of normal matter. But not the force itself; here, the splitting of the unified electroweak interaction, while not quite as violent as the inflation-inducing cataclysms of earlier epochs, was just as spontaneous and scattered. The cooling from electroweak to electromagnetism-plus-weak didn't happen all across the universe simultaneously, but rather as bubbles sparked at random places, each spreading outward.

It's like bubbles in boiling water. Except this happens at a temperature of a thousand trillion Kelvin in the first picosecond into the history of the universe as we know it. Outside the bubbles, the universe is in equilibrium. Inside the bubbles, evenness prevails as well, but in a new state. But the *boundaries* of the bubbles are different beasts altogether, and here a fully nonequilibrium state (basically by definition) occurs.

And it's there, in these exotic bubble boundaries, that all the conditions can be met within the realm of known physics: excess matter is produced, it's preserved by asymmetries in charges, and there are no competing processes there to fight against it.

Problem solved! Except that our best guess at the details of this process predict about one-billionth the expected amount of matter. Whoops.

So, yeah. After all the buildup and explanation and excruciating jargon, we don't have much. Or do we?

At first blush, this chapter so far, plus the earlier chapter discussing the earlier epochs, seems like it could be three words: "We don't know." But I hope, if I've spun this tale the way I intended, you're starting to see something interesting emerge. The further we get into the history of the universe—the older, larger, and cooler it becomes—more recognizable shapes and patterns begin to emerge from the mist.

We've gone from the Planck era, which at this stage can barely be conceived of even by hints of mathematics, into the GUT era, which has some plausible inroads that physicists have begun to explore. Then comes inflation, which Isn't That Crazy Of An Idea™, and now baryogenesis. Even though, admittedly, we don't fully grasp how matter won out over antimatter in the infant cosmos, we have a language for grappling with the problem. Weak nuclear forces, breaking of charge and parity symmetry, phase transitions, electroweak unification.

This is physics. We can do this.

Heck, even the words I'm using are beginning to change. I'm finally able to drop exponential notation and discuss temperatures (as insanely high as they are) and ages (as achingly short as they are) with familiar Greek prefixes. Instead of the barest hints of theoretical guidance, we have laboratory experiments providing clues. The fuzzy mathematics are replaced with replications of the conditions in particle colliders around the world.

Traditional cosmology books usually start with the bits we know really, really well before introducing the early-universe stuff, for good reason, because it seems like we're just making it up as we go along. But this is a book about the limits of our knowledge and the mysteries that confront us in the universe—and how we come to terms with it. And the universe prior to a picosecond in age experienced profound and fundamental changes; compared to the timescales of typical interactions, more happened in the first second of the universe than the following billions of years of cosmic history.

The first picosecond may be the dark waters where krakens lurk, but as the instants turn to full seconds, the seconds turn to minutes, the minutes to days and years, our understanding starts to crystallize. There are still plenty of mysteries out there—and don't worry, we'll get to them all—but you'll be seeing fewer *maybes* and *possiblies* and more "This is what we know happened."

Our first encounter with something far more familiar happens after the four forces are finally cleanly separated from one another. The universe is still swamped with high-energy radiation, but the chaos of the earlier epochs has simmered down into a state of matter affectionately called the "quark soup."

It's too hot in that boiling cauldron for atoms to form. Too hot for nuclei. And too hot for protons and neutrons themselves. The energies here are so extreme that even those tiny particles are ripped apart into their constituent parts.

The future history of the universe will be dominated by ever-slower and ever-less-energetic transitions. Nothing will ever reach such incredible energies again, except in isolated pockets like supernova explosions and collider experiments. The chain of events at the global scale will be dominated by pure, simple, unadulterated expansion.

As the universe grows larger, it continues to cool. Through this cooling, phases of matter can maintain their state, but eventually a thin red line is breached, and *pop*, the cosmos switches to a new form. We've already encountered a few exotic transitions as the forces of nature themselves splintered off from the crucible that proceeding them.

The phase transitions that will mark the boundaries of succeeding epochs, while still completely transformative, are much less violent. Instead of a massive cataclysm signaling the birth of a new era, they will be condensations.

Let's look at something familiar, like water, as an example before I start slinging more jargon at you. At high enough temperatures and pressures, water takes the form of a gas—water vapor. If you take a snapshot of, say, your backyard on a hot summer day and examine it at a microscopic level, you will see lots of gross bugs, but also some interesting behavior of water. Water molecules from the air will naturally condense to form a liquid on a surface, but because of the high temperature, it will immediately evaporate into the air again. The party-hearty water molecules would just love to settle down, buy a house, and start a family, but their raucous ways overwhelm the better angels of their nature.

But if you cool your yard below a certain threshold, called the dew point, the vapor-to-liquid transition will begin to overwhelm the liquid-to-vapor process, and droplets will begin to appear. Take a mass of water vapor, for instance, and start cooling it. It will remain as "water vapor but colder" for a while until the dew point is reached, when it will undergo a phase transition and become "water, as normally in a liquid."

So it's like this in the early universe, except, as you might have guessed, at much higher temperatures than you typically encounter in your backyard. From the perspective of a quark living in the first picosecond, our present-day cosmos is bone-chillingly cold and in the impossibly distant future. Keep that in mind for when I get to the chapters on the future of the cosmos from *our* perspective.

After all the craziness of the force-splitting phase transitions, we're left with our quark soup. Just like a real soup is made of chopped-up bits of larger

THE DEATH OF ANTIMATTER

things immersed in a broth, so is our universe at this time. Your molecules are made of atoms are made of electrons around a nucleus, the nucleus is made of protons and neutrons, and the protons and neutrons are each composed of three tiny little quarks glued together with—well, they're called gluons. That's what they do—they carry around the strong nuclear force.

We're pretty sure that quarks are the tiniest thing as tiny things go, so as far as we know, there's no "pre-quark stew" at earlier epochs, but be warned that picture might change.

You need to be at relatively cool temperatures to actually have a proton; otherwise, like the analogy with water, any time a trio of quarks assemble to form one, they get blasted apart by their energetic neighbors. But the expansion of the universe is inevitable, and the subsequent cooling inexorable. About a microsecond in, the first protons and neutrons begin to condense out of the early-morning fog.

It's a furious frenzy of activity. Heavy particles recondensing and reevaporating, particles and antiparticles emerging and obliterating in continuous showers of activity. But the imbalance laid down in the previous epochs persists, with normal matter having a slight edge, coming out of the fray the victor.

And it's over in a second. A single tick of the clock and our universe has gone from an incomprehensibly dense unknown state, through the splitting of the forces and a period of inflation, into a sea of familiar protons and neutrons (plus some other friends) and radiation. A state of matter, while extreme indeed, that we can recreate—briefly—in particle accelerators. So while we don't *fully* understand the physics of this epoch (as per usual), at least we can test our ideas. While "quark soup" is its cute nickname, it does have the more formal moniker of *quark-gluon plasma*, and it's something that we can cook up in labs around the world.

Even now, one second into the history of the universe, the GUT era is a relatively distant memory. Almost the entire history of the cosmos, 99.9999 percent of the time until this point, has been taken up by the formation of the first heavy particles.

And not just the first—all. Almost every single proton and neutron that we see in the universe, including the ones this book and your brain are made out of, was forged in these moments.

By the way, during this phase, the observable universe grows from about the size of the sun to about the size of a small galaxy.

The close of the first second is an important milestone in our cosmic history. The physics from here onward is even more familiar than what can be accessed deep in the hearts of particle colliders. It becomes accessible to much lower-energy devices: nuclear reactors. You could, given enough time, materials, and dedication (and access to restricted ingredients), construct in your very own backyard a device that could recreate this age of the universe. You'll also probably give yourself and your neighbors radiation poisoning, so let's just talk about it instead.

And what you would find is a nuclear maelstrom. A swarm of neutrons, protons, radiation, and neutrinos. We haven't met neutrinos yet in our story, and this isn't really their tale,[8] but something important occurs at the one-second mark concerning them.

Neutrinos themselves are nearly massless (so much so that for decades we thought they *were* massless) particles that interact with normal matter only via the weak (here we go again) force. There are billions-with-a-*b* neutrinos streaming through you right now. But they don't talk to your electrons or quarks or anybody else—except exceedingly rarely—so you don't really notice. To even get a hint of them we need literally gigantic detectors.

Neutrinos are produced in all sorts of nuclear reactions near and far, from the local power plant to the sun to distant supernovae. And the early universe. There are, at this very moment, relic neutrinos left over from this tumultuous era of the first second that have been streaming through the cosmos ever since.

Earlier than the first second, the temperatures and densities were so high that neutrinos, despite their ghostly let's-just-be-friends character, were compelled to interact with matter. But once the densities dropped, they could stream freely, liberated to live their lives as they saw fit.

Neutrinos are, therefore, one of the only ways we could *directly access* this epoch of the universe. No laboratories. No theories. Straight-up raw observational data. The downside is that these ancient neutrinos are diluted to a thin, low-energy soup here in the present-day universe, so it's incredibly challenging. But not impossible, which is important for those of you who think I'm still making all this up on the spot. This is science, folks.

Shortly after the release of the neutrinos, the *leptons* (yet another family of particles, this time comprising the light guys like electrons) fully separate out, in a similar telling of the story as the baryons (the heavy guys) but at lower, slower, energies, using more familiar physical processes.

To drive home the point that the universe is already becoming middle-aged, the era of lepton formation doesn't last a picosecond, or an attosecond, but *tens* of seconds. That's in the realm of human timescales—a couple of breaths and you've encompassed the formation of light particles! Still short, but an eternity at this epoch.

Once the temperature drops below ten billion degrees or so, as the universe ages over the course of the next few *minutes*, a remarkable process unfolds. So remarkable, in fact, that I'm about to remark upon it.

Let's set the stage: you've got a hot, dense soup of primordial particles, primarily protons and neutrons, buzzing around in a sea of high-intensity radiation and electrons, set against a backdrop of an expanding, cooling universe.

Simple question: What happens? What are the physical consequences of this scenario?

The development of the nuclear age had some serious downsides, but also an unexpected benefit. It gave us the tools we needed to answer precisely this question. Nuclear chain reactions, decay products, fusion mechanics, the whole deal. The universe at this stage is a high-powered nuclear reactor, and dang it, we know what that looks like.

It looks like a series of chain reactions: Protons and neutrons combining to form deuterons. Free neutrons decaying into protons and electrons (and neutrinos). Deuterons acquiring another neutron to generate tritium, which is radioactive but if it acquires another deuteron can generate helium-4. Free protons finding a friendly deuteron to make helium-3. Tritium meeting up with a couple of deuterium pals to make lithium-7. And so on.

These reactions can only proceed in a narrow window. Attempt to make these chains of heavier elements too early, and the sheer intensity of high-energy radiation tears everything apart, like cops at a house party. Too late, though, and the partygoers are too tired to keep dancing—the universe is too cool, too thin to sustain the nuclear fiesta.

A good solid fifteen or twenty minutes, as the cosmos cools from a billion to ten million Kelvin, is all the universe gets to form hydrogen, helium, and a little bit of lithium. While far into the future stellar furnaces will forge some additional helium and lithium, the combined might of the trillions upon trillions of stars is minuscule compared to this primordial inferno.

In other words, essentially all the hydrogen you can get your hands on, including the hydrogen literally inside your hands, and most of the helium floating around the cosmos percolated out of this epoch of *nucleosynthesis*.

I mentioned this earlier, but I'll reinforce it here: notice that I didn't say "maybe" or "we think" or "according to some random paper I found in the *Astrophysical Journal*."

This is a *prediction* of this picture of the universe, made shortly after we cracked the secrets of the nuclear code in the 1950s and '60s. The mathematics that go into this calculation have very simple results. If you happen to know the total amount of regular matter in the universe (which you can measure by counting stars and stuff), and the total amount of light (another thing that's not too hard to measure), a nuclear analysis of primordial element-making predicts relatively how much of the universe ought to be hydrogen, helium, and lithium.

It gets better: the numbers you need to know to input into the calculations don't hugely affect the results—you could not know the true amount of regular matter to within a factor of ten (that is, the real answer could be ten times smaller or ten times bigger than what you guess) and still get pretty much the same outcome in the primordial math. So you have a lot of observational wiggle room to make predictions, which are as follows: the universe ought to be about three-fourths hydrogen and one-fourth helium, with a tiny fraction of lithium.[9]

Boom. Write it down. It's a big deal. This is a moment of truth. A place where we can take this weird, complicated story of the early history of our universe and body slam it into observational reality. Our model of the infant cosmos is making a bold, unambiguous statement about our present-day circumstances. And we must ask, do we see it?

It's precisely what we find. Stars, gas clouds, galaxies. When you smear out all the *stuff* that we can see in the universe, it's three-quarters hydrogen, one-quarter helium, and a dash of everything else.

This is one of the biggest reasons we think this story is on the right track. We don't fully understand the eras that come before, but once the universe is a second or so old, it's on solid nuclear ground, ground that we've trodden on for decades. The math is complex but not intractable. The physics is difficult but not nosebleed inducing.

And it's a very straightforward, very robust prediction that agrees with the observational data.

As crazy as it sounds, it looks like this is our universe.

CHAPTER 5

BEYOND THE HORIZON

You're in the market for a house. You tell the real estate agent your budget and start the shows. Naturally, the agent has one in mind that's just a titch above your budget . . . *but I know you'll love it.* As you pull up, you get your first good look at the house. It doesn't have that much curb appeal, but it has a certain charm. The entryway is neat and tidy and organized, and the tour begins in earnest.

The agent takes you through the house, and you begin to get suspicious. A gigantic kitchen. Formal and informal rooms. A mudroom. What's a mudroom, anyway? More bathrooms than you can count on one hand. Guest rooms galore. A movie theater. A parlor. A reading room. A breakfast nook. Rooms that only serve as waiting areas for other, larger rooms.

This place is huge. Way bigger than you would have guessed from the street. The agent hasn't told you how much the seller is asking, but you're realizing that this is *way* outside your budget. There's not a single chance you'd be able to afford even the down payment, let along the mortgage.

You're not even listening to the agent blather on about the granite this and upholstered that. A single, solitary thought occupies your entire consciousness, beating your mind like a drum: *Just how big is this place, anyway?*

That, my friends, was the collective state of mind by the close of the nineteenth century and the opening decades of the twentieth. There was still heaps of angst over the nature of stars and nebulae and the causes of light spectra, and we'll get to that resolution later, but the central vexing astronomical question in the pre–Great War world was the size of our home.

Astronomers had begrudgingly come to accept the fact that our universe is uncomfortably large—parallax measurements repeated on multiple stars had hammered home the vicious point that our solar system, including the mysterious,

YOUR PLACE IN THE UNIVERSE

distant homes of the icy comets, was just a small, isolated island within the grand ocean of our galaxy. What's more, that *galaxy*—the massive collection of stars that we can and can't see with the naked eye—filled up the volume of said universe.

Or did it?

It's the natural assumption. Everywhere we look, we see stars. Yes, there might be vast gulfs of night separating the warming fires, but our universe is flooded with those fireballs. Interspersed among them are the nebulae, the vast clouds of dust and gas that serve as nesting homes or violent ends for those ferocious points of light.

The challenge with astronomy a hundred-odd years ago was that parallax is only so good. It took centuries of laborious mental exertion for the instruments and techniques to become sophisticated enough to pin down the first extrasolar distance, and once the community cracked that method, thousands upon thousands more distances were measured, verified, and cataloged. The scale of the universe was just beginning to open up before us.

But there's a limit to what we can measure with photographic plates and polished mirrors, sitting on the surface of the Earth buried under sixty miles of turbulent atmosphere. At a certain point, around a few tens of thousands of light-years, we can't get a reliable parallax measurement.

A light-year, by the way, is the distance light travels in a single (Earth) year. It's a handy unit, credited to Bessel himself for popularizing,[1] because it gets severely exhausting describing even the nearest stellar distances in "hundreds of thousands of Earth-sun distances." Instead, a typical star is, say, forty light-years away. Forty. That's much more nimble to express, so we'll keep it.

So what is a hapless astronomer to do when even the light-years stretch too far and reliable distance measurements are just a fond memory? Well, that old Newtonian idea of using the brightness of a star never quite died away, but with the stunning variety of stars on display in the sky, it always seemed like a pipe dream.

Astronomy was in need of a pioneer, and it found one in Henrietta Swan Leavitt. She had a job as a computer (back when "computer" meant a person crunching numbers by hand and not a machine that does it for you) at the Harvard College Observatory and was tasked with and/or particularly interested in the Magellanic Clouds, two cloudlike (hence the name) nebulae in the Southern Hemisphere. The clouds have been known to various cultures for millennia, but European astronomers only became aware of them in the 1400s, in part thanks to records kept during the world-girdling voyage of Magellan. Somehow his name stuck.

The Clouds contain billions of individual stars, dense stellar clumps, and smaller knots of clouds. Leavitt was particularly focusing on some specific stars within the Clouds that varied in brightness, called, well, variable stars, and here's the game she played.

If you look at a collection of stars, with some looking brighter and others dimmer, it's impossible to tell if those differences are due to variations in their true brightness (i.e., if you could examine each star close up, you would find some are blazing hot and others are dim and quiet) or whether those stars are just at different distances.

It's a mix of both in most situations—except maybe the Magellanic Clouds. They're too far away for parallax measurements, but hey, it looks like all that *stuff* is all clumped together in the same general vicinity, so maybe it's safe to assume that all the stars we'll find there ought to have about the same distance. (Very roughly, but hey, let's take things one step at a time and see how far we get.)

Of all the different kinds of variable stars (and yes, there are multiple kinds of variable stars), Leavitt was measuring the light output from a class known as Cepheids, named after the prototypical example located in the constellation Cepheus. These are giant stars with brightnesses at least a few thousand times that of the sun, but that brightness is, as you might have guessed, variable, dimming and brightening considerably over the course of a few days or weeks.

Painstakingly, Leavitt analyzed the photographic plates returned from a survey of the Clouds, comparing the same field at different nights, searching for any differences in the intensity of brightness from any of the pinpricks of light. It was mentally exhausting, repetitive work, but Leavitt excelled in dedication, and she identified nearly two *thousand* of these Cepheid creatures.

It's usually the case in astronomy that the application of large data sets reveals hidden patterns and deeper workings. When you only have a single special case or a handful of examples, it's hard to make sense of the universe. When you have a couple thousand to toss around, it's still hard, but at least you have the serum of statistics to compel nature to reveal the truth.

What Leavitt found by 1912—whether she was looking for it or not—was a remarkable pattern. If you assume that these particular Cepheids are roughly the same distance from us, then if they *look* brighter, they *are* brighter. And she must have stared in amazement at her charts and graphs when a simple relationship revealed itself to her: the brighter the Cepheid star, the longer its period between cycles of intensity.[2]

I know, that doesn't sound that amazing, but it is. This one simple relationship is about to hurtle astronomers—and us—to unimaginable scales.

Here's the deal with the deal. Measuring the parallax of these stars is essentially impossible. Measuring the brightness of these stars as they vary over time is merely insanely difficult—not impossible. And if you can connect the time variation to the *true* brightness (the astronomers among us switch to a slightly more technical term, the *luminosity*, to describe the brightness of a star if it can be measured from a fixed distance), then you can compare the true brightness to the brightness as the star appears in the sky.

And following the recommendation of our ancestral astronomers, we can do a little bit of geometry to compute a distance: there is only one number that connects true to perceived brightness. This allows us to go way, way further than crude parallax measures can take us.

Just how far? Well, that was the subject of an intense debate that I'm about to talk about. But all this history, as fun as it is to relate, is getting me hungry for some astrophysics. What is a Cepheid star, and how does it work?

Here's the basic picture, as far as we understand it. Take a giant star and wrap it in a layer of helium. Not hard, since helium is the second most abundant element in the universe. The star will heat up the helium and ionize it, ripping its electrons off. This stripped helium is a little bit opaque—light has a hard time making it through.

And so from our perspective the star looks a little dimmer, as the light from the surface is blocked by the enveloping helium. But that intense heat inflates the layer of helium, which causes it to expand, and in the inevitable process of the physics, it separates from the star and cools off.

Now more cool and collected, the prodigal electrons return to their homes, turning the helium neutral and the gas transparent. From here on Earth, we get to see the full blast of the star's intensity.

But cooling gases contract, right? So over time the helium collects near the surface, where it heats up, ionizes, and turns opaque, and—wax on, wax off—the cycle repeats again.

We're pretty sure. On the plausibility scale this gets a pretty high score, but of course it doesn't quite explain *all* the observational data.

Here's the hilarious part: *it doesn't matter*. We could be totally 100 percent clueless about how Cepheids work, with no understanding at all behind the cycles and variations. What matters, when it comes to cosmology, is that the relation between true brightness and period length holds fast. As long as the data support that relationship, we can use it to measure reliable distances. You don't have to know how your microwave oven works to heat up your ramen noodles, after all.

And now back to your regularly scheduled exploration of the universe.

Of course, we still didn't know the distance to the Magellanic Clouds. Leavitt only discovered a relative measure—the longer the period of variation in a Cepheid, the intrinsically brighter it is. And any place in the universe you can measure this variation, you can calculate a distance. But you need an anchor— a first step, using another method, to pin down the nearest Cepheid. From there you can step your way as far as your reliable brightness measures can take you.

Thankfully, a few years after Leavitt's work, a nearby Cepheid was found nestled in the Milky Way, with a confirmed distance using other methods, and the first rung of a distance ladder to the universe could be built.

But in the 1910s, not everybody was liking the taste of the Cepheid special sauce. With the benefit of hindsight and multiple decades of advancement, we now know that this brightness-period relation was on pretty solid ground, but of course at the time there were curmudgeonly skeptics—they thought there was too much fuzziness and uncertainty in the relations to use them reliably to gauge the spectacular distances under consideration. And good that those skeptics raised their contentious objections; they keep us all on our toes.

And so the debate raged on: just how big is this place, and where do we sit? These discussions came to a heated conclusion in a set of lectures on "The Scale of the Universe" held at the Smithsonian Museum of Natural History in 1920. Apparently one of the organizers had suggested an alternative topic, Einstein's theories of relativity, but this was quickly discouraged because not enough people understood it to even make for a decent debate.[3]

The two debaters, Harlow Shapley and Heber Curtis, represented the two main camps that astronomers had settled into over the past decade. It's important to relate this debate because (a) it provides a case study to set up how astronomers approach controversial issues, which will prove useful when I talk about present-day cosmological controversies, and (b) it's a fun story because Shapley and Curtis were both wrong.

In one corner, we had the Shapley camp, who thought the Cepheids were pretty spiffy and argued that our galaxy is a few hundred thousand light-years across. Plus, by noting that globular clusters (deserted clumps of old stars, or old clumps of deserted stars, take your pick) tended to be found on one side of the sky, we could imply via geometry that the sun is not at the center of the galaxy. But the galaxy, due to its great extent, filled up the entire universe, including all the mysterious spiral nebulae that dotted the evening skies.

And in the opposite corner, there was the Curtis camp, who looked askance at this Cepheid tomfoolery and insisted that our galaxy is small. The best we could do with parallax and other methods was a rough estimate of thirty thousand light-years for the diameter of the Milky Way, and no matter the direction we look, we see the same kinds of stars, implying that we're roughly in the center. But those spiral nebulae surely reside outside our own galaxy. If we assume they're the same size as the Milky Way, then they can only be extragalactic to explain their extent on the sky. And sometimes we see stars flare up—a nova—inside these nebulae. But these novae are far dimmer than those outside of nebulae. Coincidence? I think not. The spiral nebulae are "island universes," isolated homes to populations of stars far removed from us.

Two arguments, both supported by solid lines of evidence, sound reasoning, and good old-fashioned hard thinking. We can't blame the Shapleyites or Curtisans for taking the positions they did. Both sides had weaknesses in their arguments, for sure, which their opponents exploited with relish. In a way, this single lecture event encapsulated the growing frustration with the cosmos. So much was known, but it fell short of being understood. We were squeezing useful information from our telescopes and photographic plates, but no consistent story was forthcoming.

The universe was sending us mixed signals. Who could possibly sort it all out?

I won't keep you in suspense longer than I have to: it was Edwin Hubble, in the Mount Wilson Observatory, with the one-hundred-inch Hooker telescope.

Hubble found Cepheids, forty of the variable little suckers, embedded within the Andromeda Nebula, the largest (and hence thought to be closest) of the mysterious spiral nebulae. Nobody else had seen them because nobody else had a hundred-inch-wide telescope. But with Hubble at the controls of the biggest, baddest telescope ever made, sitting outside the not-yet-insanely-bright city of Los Angeles, Hubble could resolve features never before seen to humans.

Hubble published his data and necessary analysis (remember, kids, it's important to show your work) in a very readable short paper in 1925, laying out his newfound vision for the cosmos: the Milky Way is indeed large, but it is very far away from Andromeda.[4]

Hubble estimated the distance between our galaxies to be about nine hundred thousand light-years.

We now know it to be three times greater.

In a single well-written, well-argued, well-researched paper, Edwin Hubble completely repainted the portrait of our universe. And to do so, he needed a much bigger canvas.

Shapley was right in his debate a decade earlier: the sun is not at the center of the galaxy. But he was also wrong: the Milky Way is much smaller than he calculated.

Curtis was also right and also wrong: he got roughly the right size for our home galaxy but missed our location within it.

It took more data and a new round of better instruments to finally sort through the confusion, and once again the implacable data showed that the universe obeys a sort of vicious and amped-up version of Copernicus's original thoughts: We are not special. We are not at the center. And the cosmos is far, far larger than we can be reasonably comfortable with.

The Milky Way galaxy is but an island of stars—hundreds of billions of stars, but an island nonetheless—separated from our nearest neighbor by vast gulfs of almost absolutely nothing.

Just a few centuries earlier, Brahe thought it absurd to place the stellar sphere seven hundred times farther than the orbit of Saturn. And here was Hubble, sitting underneath his gigantic telescope, recording the variations in brightness from a point of light sixteen trillion Saturn-orbits away from us.

The Andromeda Nebula was no banal collection of gas and dust with a dash of errant stars loosely tossed within it. It was a galaxy in its own right, one of an uncountably large multitude, spread throughout the achingly large cosmos the way a child might leave toys scattered in a room.

Nobody was really thrilled at the result. Hubble demonstrated that even the "large" universe as advocated by Shapley was too small. Let that sink in: the Cepheids used in Edwin's analysis were farther away than even the *farthest* possible thing a reasonable person could argue existed.

And Andromeda wasn't alone. Now that Cepheids could be reliably used to measure extragalactic distances (and "extragalactic" was now a thing), many other distances were pegged to other nebulae-*cum*-galaxies. Andromeda has pride of place of being the first, but it was far from the last.

But the fun didn't end there. Oh, no, child.

Not satisfied with simply rescaling our universal yardsticks and taking the first crack at a truly cosmological birds-eye view of our home, Hubble took it one step further and completely revolutionized our understanding of the *dynamics* of the universe writ large.

Remember how much intellectual inertia had to be overcome to finally conclude that the heavens were as violent, chaotic, and messy as the physics here on dear old Earth? Centuries, that's how long. In those intervening generations, scientists played a sort of slippery rhetorical bait and switch. Sure, stars can blow up, change their brightness, and even scoot around. Whatever. But the *universe* is eternal and never changing.

Individuals may come and go, but life goes on, forever into the past and forever into the future. It's the way it has been, the way it is, and the way it always will be.

Academics replaced the fixity of the firmament for a constancy of the cosmos. It's the same thinking, just with a few zeros tacked onto the end of all the numbers.

And yes, Hubble had to go and pop that bubble too.

In 1929, four years after his landmark presentation on the true distances available in our universe, he published another, highly readable paper in which he reported an interesting relationship between the distance to a galaxy and, of all things, its speed.[5]

Remember the spectral Doppler technique used to measure the motions of stars? (If you don't, you weren't really paying attention to chapter 3, were you?) The beauty of that method is its brutal universality. Find a star, measure its speed toward or away from us. Boom, done. Move onto the next.

Find a nebula, measure the light. Recognize any emission or absorption lines from your favorite element? Is the fingerprint correct but shifted left or right from its Earthbound counterpart? Congratulations, you've measured the speed of that nebula—even different parts of it!

Is that "nebula" really a gigantic galaxy, home to hundreds of billions of stars, as big as or bigger than the Milky Way, located hundreds of thousands of light-years away? Who cares? It's emitting light, so we can take a spectrum, either from individual stars, if we're lucky enough to resolve them, or from the generic glow of the galaxy itself.

Recognize the elements? Fingerprint shifted? You've measured the velocity.

If it's shifted toward the blue, it's coming toward us. If it's shifted toward the red, it's moving away.

Of course, just like with stars, this technique only returns the *radial* speed, the speed along our line of sight to the object. In other words, the speed along the in-out direction. Who knows what the up-down or left-right speed is. But hey, it's something, and we'll take it.

Specifically, Hubble took it, twenty-four times. It was the best he could do through painstaking (Have I used that word enough to describe the procedure of astronomical observation? No, I haven't.) work collecting, measuring, recollecting, and remeasuring the tediously extracted distance measurements from the Cepheids, then matching those distances to velocity measurements taken by himself and previous astronomers.

What he found was simple, surprising, and essentially inarguable: galaxies, on average, are receding away from us. And the farther away a galaxy sits from us, the greater its redshift, implying the faster it's receding from us. *And* this relationship is linear: double the distance, double the speed. Quadruple the distance, quadruple the speed.

Hubble could even provide a number for the recession rate: five hundred kilometers per second per megaparsec.

As we'll soon see, the numbers and counting systems we're accustomed to quickly become too cumbersome to have any utility, a fact that astronomers quickly realized probably just after the phrase *astronomically large* came into circulation. The standard go-to is the light-year, which as you recall is the distance that light can travel in a year—5,878,499,810,000 miles, for the curious.

But astronomers typically use a different measure: the *parsec*. Why? I honestly don't know—maybe because it sounds cooler, and maybe because the term *light-year* was initially used more for the benefit of the popular imagination and not for use by Serious Astronomers. But anyway, parsec is short for *parallel arc second* and comes in handy when measuring astronomically large distances. If you hold a pencil in front of your face while closing each eye individually, the pencil will appear to wiggle back and forth relative to the distant background. If you hold the pencil farther away, it will wiggle less.

If you know how far apart your eyes are, and you measure the amount of wiggle, and you know about trigonometry, you can calculate a distance. This is the parallax method.

That isn't so useful for interstellar measurements, so instead of alternating eyes, astronomers alternate seasons, repeating measurements when the Earth is at opposite sides of its orbit around the sun. This is the precise technique that Brahe used to cast doubt on the sun-centered model of the universe and Bessel used to give him a run for his money. And here comes the definition: a parsec is the distance an object has to be at in order to wiggle by one arc second (1/360th of a degree, for those not nautically inclined) when observed six months apart.

One parsec is roughly three and a quarter light-years, and Proxima Centauri, our nearest neighboring star, happens to be around one parsec away. Hmmm, I wonder why that definition was chosen?

And just as your computer can have megabytes and gigabytes, representing a million and a billion bytes respectively, distances can have megaparsecs and gigaparsecs. Start tossing those kinds of words around, and people won't even know you're not a real astronomer.

As Hubble's result implies, we're going to use those: the Andromeda Galaxy, née Nebula, our nearest major neighbor, is about three-fourths of a megaparsec away from us.

So here's what Hubble's calculation reveals: for every megaparsec you get away from the Milky Way, objects at that distance are receding from us an additional five hundred kilometers per second.

At some level, it's not surprising to know that galaxies zip and zoom around. People do it, planets do it, stars do it. Why not the largest collection of all? But something fishy was going on with Hubble's measurements. In the peculiar jargon of astronomy, the individual velocity of any particular galaxy is called its *peculiar velocity*. So fine, the peculiar velocities of galaxies aren't zero. I suppose it's a little bit of a big pill to swallow to contemplate the motion of these humongous cosmic assemblages, but it's one we can take.

It's the average motion that's troubling. There appeared to Hubble to be a separate, common (dare I say universal?) motion to the galaxies around us.

And specifically, away from us, at the very precise rate of five hundred kilometers per second for every megaparsec in distance. This was certainly something new and potentially troubling. Given the raw data, how are we to possibly interpret this result?

The revelation of two Hubbles. *Left,* a modern view of the Andromeda "Nebula," which Edwin Hubble conclusively demonstrated was really, really far away. (Image courtesy of NASA / JPL-Caltech.) *Right,* a long stare with the Hubble Space Telescope at a small patch of sky—equivalent to the small square next to the Moon—reveals a universe infested with these beasts.
(Images courtesy of NASA / ESA.)

Option 1: A conspiracy. Galaxies move around randomly, and they *just so happen* to have the right velocities so that a galaxy, say, twice as far from the Earth is moving twice as fast as its nearer cousin. And they're all moving away from us. Maybe we're the center of the universe, and we're somehow repulsive?

Option 2: An illusion. Astrophysicist Fritz Zwicky, who really knew how to rock a bolo tie and whom we will meet again later, tossed this idea into the discussion. Maybe light just gets tired, like an out-of-shape guy trying to run a marathon. Who knows what the mechanism is, or what the physical implications might be? But in this case perhaps the general redshifting isn't an indication of motion but a loss of energy. Redder light is less energetic than bluer light, so this hypothesis is still quite capable of fitting all of Hubble's data. The recession is a fake; it really isn't motion but an artifact of our poor understanding of physics.[6]

Option 3: We live in an expanding universe.

You and I know that the answer is door number 3, but it's not easy to flesh out that deceptively simple but radical statement in a single paragraph, so it gets its own section.

One of the most amazing aspects of this saga of the 1920s is how quickly the resolution came. When Copernicus and Kepler first proposed a sun-centered universe, it took another couple of generations before Newton could offer a unifying theoretical theme, a coordinating force ("universal gravity") that could sufficiently explain the motions and models offered by earlier thinkers.

But in this case, the theoretical basis for Hubble's observations was happening simultaneously—and had even been anticipated before he got the result! That theoretical basis is general relativity, and if you've been wondering when dear old Albert would get fully introduced into the story, well, here he is.

Einstein had long been attracted to the problem of gravity (get it?), and especially to Newton's troubled comments that he didn't fully understand *why* his relationship holds between massive bodies in the solar system, but that it nonetheless works, so it must have some utility.

I'll have to save a full and proper treatment of Einstein's work and legacy to another book—after all, he did basically found a few distinct branches of modern physics—so here's the short and sweet version (as short and sweet as I can make it[7]).

General relativity is Einstein's magnum opus, a completely radical take on the underpinnings of gravity, explaining it not as a *force* per se but as an *effect*. Instead of gravity being an instantaneous, invisible communication between massive objects, the effects of gravity in Einstein's picture are actually the consequence of a relationship between mass and space-time itself.

Right, space-time. Not space *and* time. Not space *or* time. Space-time. A single, unified thing. We don't live in a three-dimensional world; we act our little plays in a four-dimensional stage, with our three spatial dimensions (up-down, left-right, back-forth) and a fourth dimension of time (past-future). This was the revelation of special relativity, one of Einstein's first forays into reshaping our worldview in 1905, but it was another scientist (and Einstein's former teacher), Hermann Minkowski, who later made the leap into unifying the dimensions.[8]

I mention this because you're going to be seeing the word *space-time* a lot, and I need to explain what it really is: It's a ruler. It's a way of measuring distances (which doesn't sound like a big deal) in both space and time (which does sound like a big deal). If we agree to meet for coffee at four in the afternoon, the total description of the location ("coffee shop at 4:00 p.m.") is a precise *event*, and the construction of space-time allows me to compute how far I have to travel to get to that event: I have to move, say, 3.5 miles west, *and* I have to wait until it's two hours into the future before the chitchat over lattes can occur.

You can imagine space-time as a grid (in four dimensions, so good luck), marking out locations throughout the entire universe, where "entire" also means the distant past and future.

It's a dance floor. The particles, forces, fields, and energies that populate our universe are the dancers, twisting and twirling away in complicated rhythms and beats. But the space-time floor stays fixed.

Or at least it did in the decade between 1905 and 1915, when Einstein brain-birthed general relativity. The special version recognized certain rules that applied throughout the universe: the speed of light is the fastest anything can travel; moving clocks run slow; different observers will disagree about lengths and intervals; mass and energy are two sides of the same coin; and so on. To make special relativity happen, Einstein had to chuck out the Newtonian framework of gravity; it simply wasn't compatible with the new set of rules.

For example, what if the sun vanished? Light takes eight minutes to make the leap across the vacuum from our star to our squinting eyes, so we wouldn't know the sun had disappeared until eight minutes after the event occurred. But would the orbit of the Earth change in those eight minutes? Would the inertia of the Earth "know" instantaneously, while light lagged behind?

Einstein didn't think so (and to be fair, had Newton been aware of the issue, he probably wouldn't think so either). Rather, something had to *carry* inertia, to take it from place to place in the universe—to connect motions across the vastness between us.

So what infrastructure exists that permeates the cosmos, allowing all the particles, forces, fields, and energies to interact with each other?

Bingo: space-time.

It turns out the dance floor isn't as solid as we thought it was. It doesn't just stay there, a rigid platform for the drama of the universe to play out on. Using a mathematical tool kit developed in the nineteenth century by Bernhard Riemann, Einstein was able to formulate a view of the universe where the floor—space-time itself—bends, warps, flexes, and curves.

Imagine yourself gliding down the floor in a smooth-as-butter waltz (or a sassy hip-wiggling salsa, if that's more your flavor), but the floor is a trampoline. Your very presence bends the floor underneath you, as it does to all the other dancers. Negotiating the limited space in the crowded room is a tricky thing. If you even get near the other dancers, their dance-floor depressions alter your course, veering you away from your intended movement.

And that, my friends, is the Einsteinian picture of gravity. Except in four dimensions. Sorry, folks, but analogies can only take us so far in a space-time world. The presence of matter (and energy!) distorts space-time around the object, bending it. Any other matter (and energy!) encountering that object will have its motion disturbed by that deformation in space-time. That is our gravitational experience.

This picture answers the vanishing-sun riddle: if the sun were to *poof* out of existence, it would take a while for space-time to "relax" back to its flat state and for the Earth to be released from the gravitational grip of the sun, so our planet would be flung out like a spinning rock cut from its string at the same time that the familiar light in the sky would wink out of existence.

So special relativity gave the world a language of space-time, and general relativity taught us that space-time itself is a dynamic, living, breathing, physical object. It's still a ruler—it's still very good at measuring intervals between events—but that ruler can stretch, flex, and bend.

The game of general relativity is then pretty straightforward. In words, that is; in the mathematics, it's insanely complicated. Gravitational interactions are formulated as a set of ten interconnected nonlinear equations, with one side of the problem describing all the possible ways that space-time can bend, flex, and twist and the other side of the problem describing all the ways the matter and energy can bunch together flow, and twist.

So in most cases you take a given physical system—say, the solar system. You count up all the matter and energy sources you care about (e.g., the sun and planets), you turn the GR crank, and out pops a configuration for space-time. *Then* you apply what's called an equation of motion to, well, figure out how the objects ought to move. And boom, you've got some dynamics.

What does this game have to do with the universe? Well, the universe can be considered as a single, physical system. It contains a certain arrangement of particles, forces, fields, and energies. All that *stuff*, when studied from the ultimate long view, smoothed out across the entire universe, will bend the cosmos at those very largest scales. In other words, the contents of our universe will bend creation itself, and that bending will influence motion at the largest scales—say, by sending galaxies flying away from each other.

Einstein was the first to apply his tools of general relativity to questions of cosmology in 1917, just a couple of years after formulating the methods in the first place.[9] To be fair, he was only one of only a handful of people who actually understood how to do it, so it might as well have been him.

But 1917 was twelve years before Hubble's result, and Einstein assumed, just like everybody assumed, that we live in a static, eternal universe. The firmament was fixed, just like the ancients thought, but the word "firmament" had taken on a much larger definition.

What's interesting is that general relativity didn't automatically predict a static universe—left to their own devices, the equations naturally suggest a *dynamic* cosmos, one that's inclined to expand or contract but not stay still. Well, that wasn't going to work. No way was the whole entire *universe* moving around. So to develop a halfway decent model of the static universe as Einstein and everybody else knew it, he had to plug a somewhat awkward "bonus term" (not his words) into the equations.[10]

It was a perfectly reasonable decision to make at the time, and nobody gave him any gruff for it. But imagine for a moment a world where Einstein didn't feel compelled to fit the known data—where he let the simplest possible expression of general relativity *predict* what the universe ought to be like. He would have come out with the dynamic universe a full decade before observations would have backed him up on his bold claim.

Man, he could have been famous.

Thankfully, Einstein wasn't the only one thinking of cosmological problems and using the general relativity tool kit to think those thoughts. Many other theorists toyed and tinkered with Einstein's equations, including Willem de Sitter (a Dutchman with a pointy beard), Alexander Friedmann (a Russian with a caterpillar mustache), George Lemaître (a clean-shaven Belgian Catholic priest), Bob Robertson (an American with the tiniest mustache you've ever seen), and Arthur Walker (an Englishman with no beard).

Their story of attempting to use general relativity to describe the whole universe is long, intricate, and intertwining. The most important point is that it demonstrates that no matter your choice of facial hair arrangement, you too can be a theoretical physicist.

It also provided the mathematical framework for modern cosmology. In an expanding universe, the galaxies only *appear* to be physically rocketing away from each other. In fact, the fabric of space-time itself stretches like pizza dough, causing every galaxy to separate from every other galaxy (ahem, on average, and that caveat will have some interesting consequences for later chapters). The redshift that Hubble noted thus is due to not the motion of galaxies but the stretching of space-time. Make no mistake; the galaxies really are getting farther away from us, but not on their own agenda—the increasing gulfs between us are like the tectonic spreading of oceans between continents.

As light travels from a distant galaxy to, say, the aperture of a hundred-inch telescope, the expansion of the universe stretches out the light. Farther galaxy = more intervening universe = greater stretching = more redshift. Perfectly matching Hubble's results. No intergalactic conspiracy needed.

The gist is that by the time Hubble announced his findings to the world in 1929, he had done his homework and knew enough to name-drop Willem de Sitter's work in his paper as a possible explanation for the results—that we live in an expanding universe.

Modern cosmology can thus trace its lineage to two founding fathers. One was Einstein himself, the theorist's theorist, who was profoundly unconcerned with the experimental results of tests of his theories—what else could they possibly find but that he was correct? A solid mathematical argument could sway his thinking, but *data* would only serve to validate his reasoning. Indeed, the one time he toed the observational line by assuming a static universe, he modified his equations in a move he would later call his "greatest blunder."

The other was the pipe-smoking, eagle-eyed observer of the cosmos, Edwin Hubble. The champion of solid analysis, good statistics, and simple but powerful writing, Hubble was extremely cautious about offering theoretical explanations for the amazing results he achieved—it was only in the closing paragraph of his landmark distance-velocity paper that he remarked that the results *might* be explained by an expanding universe.

In other times, both past and present, observers and theorists are often at odds, either leapfrogging each other, sneering at the opposite camp in the rearview mirror, or straight-out devolving into fistfights.

But something magical happened in the decades surrounding the world wars—for a short period of time (cosmologically speaking, and also in the sense of the timescales of human achievement and growth of understanding), those who collect data and those who try to explain data were in almost perfect lockstep, sometimes even publishing papers side by side in the same journal issue.

The universe was starting to come into focus. For a brief moment, everything felt *good*.

CHAPTER 6

BATHED IN RADIANCE

So far in our tale we've been following two threads. One has been about humanity's general confusion when it comes to the goings-on of the night sky, and our attempts—usually feeble, but occasionally breathtaking in scope—to measure and understand what's going on up there. The other thread has been a biography of the universe itself as we currently (don't) understand it, starting in the black box of the Planck epoch and proceeding through the splitting of the forces, the incredible dynamics of inflation, and the rise of matter over antimatter.

It's time for these two threads to—briefly—meet. The next major event in the history of our cosmos is a watershed transformation, a clear dividing line between the exotic forces and energies of its youth, and the beginnings of a structure that we will eventually grow to call familiar.

In the timeline of our more recent history, the expansion of the universe had just been uncovered, and in the coming decades, debates would swell over how best to interpret Hubble's stunning results. But in the 1960s, a key observation would be made—a simple collection of data that cemented our modern picture of the grand history of our universe: the big bang model.

By the time our universe was twenty minutes old, it had already experienced the most dramatic phase changes it would ever experience. Imagine, if you will, that in the instant after your birth you immediately experience growth spurts, puberty, and the onset of middle age—complete with graying hair—before the doctors have even cut the umbilical cord. While in terms of the linear passage of time, the universe has a long, long future ahead of it, by the end of the nucleosynthesis era, it had already experienced the most exciting things that could happen to it. It was then doomed to a relative retirement and decline for the rest of its days.

Corresponding with the change in character of the universe is a change in important timescales. When the universe was dominated by exotic merged forces like the electroweak interaction, the operational physics was set by the speed of that dominant player, and phases would begin and end in the blink of a cosmic eye. But once the protons, neutrons, and electrons that we're familiar with in our daily lives were finally manufactured, the universe took its first slide into the slow lane of life.

It was inevitable, really. Continued cosmic expansion means continued cooling. At lower densities and lower temperatures, more familiar physical interactions take precedence. Strong and weak nuclear have each had their turn, and now and for the next few hundred thousand years, it's time for electromagnetism to take charge (when my editor left a note saying, "I see what you did there," I realized that this was perhaps the only unintentional pun in the entire book).

Indeed, if you add up all the stuff remaining in the hot early universe, it's mostly composed of photons, the ubiquitous carrier of the electromagnetic force. The cosmos at this stage is a proper plasma, the same state of matter you would find in a lightning bolt or the interior of the sun (please don't go personally looking for this; just take my word for it).

In a plasma you've got protons, neutrons, electrons, and photons, all bouncing around together angrily. And some of the protons and neutrons have hooked up to form helium or lithium, and the energy of the surrounding soup is now too feeble to break them apart.

But the battle between matter and antimatter was a Pyrrhic victory—only one proton per billion survived the great primordial war, leaving the numbers of normal matter severely underpopulated. So the plasma was strongly dominated by the photons, the light. Within this opaque energetic soup, atoms (yes, finally, *atoms*) would try to form when a stray electron would get caught in an orbital around a proton or other nucleus. But as soon as the bond would form, a home-wrecking photon would slam in to destroy the newfound relationship, slapping the electron away and back into the mix of singles.

Nothing could withstand the domination of the photons. Strong nuclear was too short-range, and in the expanded cosmos, it could only find its influence confined to the nuclei. Weak nuclear, as exotic and essential as it can be, was always a pushover. And gravity? By far the weakest of the forces, billions upon billions (upon a few more billions) of times weaker than even the so-called weak force, it was hopeless against these interactions.

But gravity did have one thing going for it—it was playing the long game, fighting with guerilla tactics, laying traps and ambushes so that the tyrannical photons would eventually sow the seeds of their own downfall and forever be relegated to cosmic irrelevance.

Gravity won its ultimate victory by fighting dirty. It couldn't hope to free the helpless atoms from the relentless electromagnetic onslaught in one-to-one battles. But Einstein, de Sitter, Friedmann, Lemaître, and all the others had discovered that gravity, and gravity alone, was sufficient to describe the dynamics of the universe at the very largest scales. While other forces may govern small interactions (like the formation of an atom an hour into the history of the cosmos or the rhythm of your heartbeat billions of years later), gravity is the only force that operates at infinite range and affects *all* things, regardless of size, shape . . . or electric charge.

That's the key. Photons could win individual battles, but the war was over before electromagnetism even started marshaling its forces. That's because the universe is, on balance, electrically neutral. For every negative charge, there's a corresponding positive charge out there, somewhere. So on average, any large-scale electromagnetic interactions simply cancel out. There's no coordination, no grand strategy, no overarching plan.

And gravity had a plan. Gravity was driving the expansion of the universe—that's the ultimate lesson of general relativity—and expansion makes the density drop for everyone. For matter, it's a simple cubic relationship. Put one particle in a box, you have a density of . . . one particle per box. Expand the box by doubling each side, you now have the equivalent of $2 \times 2 \times 2 = 2^3$ boxes, so your density has dropped to one particle per eight boxes.

This goes for matter and radiation in equal measure, so at first blush it seems like a wash. But photons pick up an extra interaction from the expansion of the universe. They get elongated by the stretching of space-time itself—as the universe grows, their wavelengths grow longer as well. In other words, the light redshifts, which is what Hubble observed when he examined the light from distant galaxies in the 1920s.

And here's the kicker: redshifted light has less energy.

It's happening in our universe now, and it happened in the universe long ago. As the universe aged, the photons not only diluted, but also lost energy. It was an endurance race between matter and radiation: who could outlive the other in the inexorably evolving cosmos? This was gravity's wicked plan all along; the radiation was simply born to lose.

Because radiation had such overwhelming numbers, this was indeed a long struggle, the longest such war for survival that the universe had yet seen. But gravity is gravity is gravity: quiet and unassuming, but relentless.

The universe expanded and cooled. The densities of matter and radiation dropped. The primordial plasma lost its ferocity. Year by year, the photons grew tired, their influence over matter diminishing.

And 380,000 years into the history of the cosmos, when our observable universe was about one-billionth its present volume, radiation gave up the fight for good.

The first atoms were born.

The name "big bang" was coined by the sharp-minded and sharper-tongued astronomer Fred Hoyle, who didn't exactly take a shine to this expanding-universe business.[1]

Hoyle had a good point—Edwin Hubble's 1929 observations of a general redshifting of light from distant galaxies only *suggested* an expanding universe. But the "conspiracy" option, where we are the literal center of the universe with all the galaxies literally flying away from us in a predetermined pattern in order to reproduce a straight relationship between distance and redshift, was never really under serious consideration.

Why? Because Copernicus, that's why. Hundreds of years ago, "Hey, folks, maybe the universe isn't focused on us, just saying" was a radical, thought-provoking notion worthy of much narrowing of the eyes and muttering under the breath. But the eventual success of that (initially flawed) picture, plus decade after decade of the universe rubbing our noses in it—proving over and over that the fantastic energies and vast scales *really* don't care about us—led, by the twentieth century, to a much more cautious generation of astronomers.

For them, it was *preferred* to assume that we're not special. Better safe than sorry, I suppose. This central conceit of modern cosmological thinking goes by a few names, like the Copernican principle or the mediocrity principle, and we'll return to the topic in sobering discussion later.

What about "tired light," the concept that it's not expanding space that's sapping the energy from light, shifting it to redder hues, but simply that light loses energy as it travels? One challenge to this idea is that in order to make light grow tired, you must have that light interact with some sort of substance sprinkled in the intergalactic gulfs—say, magical redshifting pixie dust, purely for example. Since the light will bounce off that magical redshifting pixie dust, the light must also scatter. So images of distant galaxies must be slightly fuzzier than closer ones, because their light has had more interactions with the magical redshifting pixie dust. Plus, that same magical redshifting pixie dust must be sprinkled *inside* our own galaxy too, so stars on the far side of the Milky Way should be redder and fuzzier than our closer neighbors.

The instruments of the first half of the twentieth century didn't have quite the measuring sophistication to conclusively rule out tired light, but the concept never really caught on. There were no known physical mechanisms for making light tired, and it conflicted with everything else we knew about the photons among us. Even Fritz Zwicky, the bolo-rocking astrophysicist who cooked up the idea, tossed as many varieties of models into his paper as he could think of (and crossed a few off the list in the very same paper), in the spirit of "Let's make sure we don't leave any stone unturned before we jump into an expanding universe."

For those of you with highly skeptical hearts, don't fret. More modern astronomers with instruments of sufficient sophistication have indeed followed up on these lines of thinking and found them to be less than fruitful. Tired light is a tired idea.[2]

Expanding universe it is, then. But perhaps not necessarily the big bang "primordial atom" as cooked up by the Catholic priest Lemaître in his application of relativity to the universe. After all, the cosmos having an "origin" did seem bit too close to Genesis for some, and didn't we move past all this using-the-Bible-to-support-our-arguments line of thinking since the days of Kepler?

Enter Fred Hoyle, an amazingly brilliant astronomer who, as far as I can tell, decided to take up the mantle of Curmudgeon Superior from Galileo and seemed to openly work against his own best interests, burning bridges faster than he could build them. He led vital work into the nature of how stars function, but in this story of the universe's story, he serves as the devil's advocate against the consensus growing around Hubble and Einstein's cosmological offspring.

And, like before, I have a soft spot in my heart for the die-hard skeptics in history, even when they become so cantankerous that nobody invites them to any parties. They're annoying, but oh so useful.

The ultimate too-cool-for-school kind of guy, Hoyle often took the opposite position to whatever was popular with his fellow scientists. I must say this was an awesome tactic, because (a) science needs healthy debate and skepticism to survive, and (b) he was smart enough for it to pay off most of the time.

But if a cosmic conspiracy and tired light were off the table for cosmological consideration, what possible alternative explanation was there? You couldn't argue with the data—the results of Hubble and company were too squeaky-clean for any charges of shenanigans. But you could always argue against the theory. Not general relativity itself—by the 1930s, Einstein's theory had already trounced any other potential challengers to the title of Explainer of Gravity—but there was one little crack that Hoyle identified. A small one, but big enough for him to drive a wedge into it and force the scientific community to hold it, take a breath, are you sure?, before jumping off the cliff.

The big mental hurdle that you have to leap, the metaphysical pill you have to swallow, the elephant in the room that you have to address if you want to take this big bang picture seriously is that the universe has, fundamentally, finite age. It has a *beginning*. There is a specific moment, in the countable past, when the universe switched from not existing to actually existing.

If you're religiously minded, that's not such a big deal. But Hoyle wasn't arguing against the so-called big bang (and even though he didn't intend the tag to be derisive, given his cantankerous nature I can't help but see the corner of his lips curl when he coined the phrase during a BBC radio show in 1949) theory on religious grounds. Far from it. "Everything we see in the universe came from, well, somewhere" isn't the most scientific of statements, but on its face it's not malignant.

Instead, Hoyle challenged the cosmic establishment to go all the way to the finish line when they insist that the universe doesn't care about us. If you're going to elevate Copernicus such that his name gets stuck in front of the word "principle," the thinking goes, then you need to finish what you started.

We are not the center of the universe. We are not special. We do not have a special vantage point on the heavens—our view is, statistically, just like anybody else's. Ergo, from our perspective it looks pretty much the same in every direction. In the jargon, our universe is *isotropic*.

You can take it one step further and assert/assume that on average, at the largest scales, the universe is generally the same from place to place. It is *homogeneous*, like the milk you buy from the store. Therefore, *nobody* is the center. There is no special location in the universe that is wildly different or elevated or distinct from any other. Again, I'm repeating the phrase *on average, at the largest scales* because this is cosmology: you can only think big about these kinds of questions.

These two ideas combined, that the universe is both isotropic and homogeneous, form the backbone of general relativity's insights into the cosmos, and hence they are often referred to as the *cosmological principle*. They are the basic assumptions needed to simplify Einstein's nasty equations enough that you can get work done with the mathematics, *and* they serve as fundamental statements about the nature of our universe and our role in it.

Hoyle and colleagues rattled the cages: you want a *boring* universe, where nothing is very different from place to place? That's fine, that's great, that's wonderful. So then shouldn't the universe be pretty much the same from time to time as well? In other words, shouldn't our cosmos be the same through space and time together, like in this concept of *space-time* that everyone is so excited about?

The refutation to the big bang's cosmological principle was a *perfect* cosmological principle that rested on the assumption that the universe is eternal and unchanging, that it is indeed homogeneous, through both the vastness of space *and* the deepness of time. This was the default position just a few decades earlier, before Hubble astounded the world, so why let his results spoil the fun?

This might have ended up just empty words, but like I said, Hoyle had chops. Together with some colleagues he formulated an attractive alternative to the big bang—the steady-state model.[3] In this picture, requiring only a small and innocuous alteration to the equations of general relativity, matter is continuously created in the universe, with the rate of creation matching the outward expansion. Thus as the universe continues to grow fatter, its density remains constant—there are always new partygoers joining the big bash as the room gets bigger.

Steady-state cosmology fit the data just fine. Arguments that it seemed too absurd to have matter popping into existence all the time ("Where does your stuff come from?") were met with sharp rejoinders—the big bang model also posited the spontaneous creation of matter ("Where does *your* stuff come from, pal?"). It simply stretched the instant, fiery explosion of matter into a long, drawn-out slow burn. A simmer rather than a boil.

The match was set, between the perfect cosmology of the steady-state picture and the finite-aged cosmos of the big bang. And through the late 1940s and into the 1950s, there was no clear winner.

After 380,000 years of waiting, electrons could finally join their hadronic cousins and form the first atoms. Before this time, the radiation had already diluted to the point that matter was the dominant player, but still it fought its losing, helpless battle, preventing the formation of atoms. Finally, though, it called it quits; matter and radiation would never affect each other on cosmic scales again.

This remarkable event would have just flashed by in a haze known to us only dimly via equations and simulations, like all the other major transitions before it, except that the universe was now, for the first time, transparent. Before the formation of atoms, the universe was filled with hot, dense plasma. Just as the radiation prevented the atoms from forming stable long-term bonds, the thick dilution of matter prevented the radiation from traveling freely. A photon would attempt to make a great leap at light speed, only to run smack-dab into a klutzing electron.

But now that neutral hydrogen and helium had formed, deliciously transparent, light had room to move. Over a relatively brief window of time, about ten thousand years or so, the fog of the primordial universe lifted, and a more recognizable, more *clear* universe became the norm.

For obscure historical reasons, physicists refer to this event by the name *recombination*, as if this were the second time that atoms got together in the universe, which it wasn't, unless you count being squished into an exotic quark-gluon plasma as "together." I personally prefer *photon decoupling* or, less formally, *the best fireworks show ever*.

The light emitted was literally white-hot, corresponding to a blackbody temperature of about three thousand kelvin, about half the temperature of the surface of the sun.

I know, I know. Blackbody temperature? It's perhaps one of the most confusing terms in all of physics (and that's saying something). It comes from the devices used in the nineteenth century to study the radiation emitted from as-black-as-possible objects, objects that drank in as much of the surrounding radiation as possible and were at a fixed temperature. Perhaps a more descriptive term is *thermal radiation*, or maybe even *warm and/or hot stuff radiation*.

All stuff gives off radiation of some form. All that wiggling, jiggling, and rotating at the molecular level releases some of that energy in the form of light. Since some wiggles and jiggles are bigger or smaller than other wiggles and jiggles, the radiation emitted covers a broad spectrum, with a distinct peak depending on the temperature.

For example, you. At a temperature of ninety-eight degrees Fahrenheit, you are emitting all sorts of radiation, most of it in the infrared—hence why infrared goggles are so handy for seeing people in the dark. But you're also giving off a little bit of microwaves (enough to be detectable by a standard household satellite dish) and even visible light (not enough to be seen, but it's there).

The cooler an object is, the longer the wavelength of the majority of light it gives off. The hotter, the shorter. The full description of blackbody (aka thermal) radiation was cracked by Max Planck, a name we already encountered as we tried to come up with a numbering system to describe the earliest moments of the universe. In the process of describing blackbody radiation, he also inadvertently invented quantum mechanics, but that's a story for another chapter.

At the moment of the separation between radiation and matter 380,000 years into the history of our universe, the cosmos was in almost perfect equilibrium. Radiation and matter were bouncing around ferociously, and those countless interactions created an essentially ideal blackbody scenario. Thus when the light was finally released, it carried that imprint, perfectly mimicking a laboratory device at a temperature of three thousand kelvin.

That primordial light permeated the cosmos. Truly for the first time, the densities had dropped so much that it could travel for countless light-years before interacting with a stray bit of matter. It soaked the universe but was no longer a part of it. And it was *bright*, like having the surface of the sun surrounding you on all sides. Indeed, this sudden release of radiation generated more photons than all the stars will produce, ever, in the entire future history of the cosmos.

But that event was a long time ago. We aren't bathed in white-hot radiation from the early universe. What happened? The quiet but inexorable expansion of the universe happened. Gravity didn't just win; it rubbed radiation's nose in it. With the continued expansion, the radiation was stretched and stretched, redshifted down just like any other long-distance photon in the universe. The primordial light was still there, bathing the sky, but no longer in the visible range of the human eye.

By the late 1950s, steady state was starting to look a little unsteady. Newly developed radio telescopes were beginning to peer into the deep cosmos, and their initial results revealed an unexpectedly high number of intense radio sources, at great distances, and a relative radio silence nearby. That's a hint—only a hint—that the universe may be the same in space, but different in time. If light takes a certain amount of time to travel from place to place, then the further we look, the deeper we see into the past. If these radio emitters, known as quasars, are out there but not around here, then that means they were more common in the past. Perhaps then the universe has changed character with time.

Around the same time that Hoyle was pooh-poohing the big bang, other scientists were working out the full physical implications of such a radical universe. Four in particular—Ralph Alpher, Robert Herman, George Gamov, and Robert Dicke—semi-independently came to a remarkable conclusion. If the universe were smaller in the past, then it must have been hotter. Eventually, at some distant time, it should have been so small, dense, and hot that it was a plasma.

But at a specific time, a switch would have flipped, juuuust as the universe cooled enough to the right amount, and the cosmos would have gone from plasma to not-plasma. And that radiation has a calculable temperature, based on our knowledge of plasma physics (which was all the rage at the time), but that temperature would be reduced by the present epoch.

They initially calculated a temperature of a bare few degrees above absolute zero—apparently our present-day universe is indeed very cold—corresponding to a peak blackbody wavelength firmly in the microwave band. Additionally, we should be completely soaked in this radiation; if our universe is truly homogeneous, it should fill out the sky equally in all directions, with nary a deviation in sight.

So all they needed to do was design, build, test, and operate a microwave antenna and search for this "cosmic microwave background," and they'd be set.

At the same time, two engineers for Bell Labs, Arno Penzias and Robert Wilson, who, bless their hearts, knew absolutely nothing about cosmology, were designing, building, testing, and operating a microwave antenna for their own industrial purposes.

It was the "testing" part that was giving them trouble. It was the first time in human history that someone had attempted to construct such a microwave detector, so I can't fault them for making things up as they went along. They built everything perfectly, but try as they might, they couldn't get rid of a constant background hiss from their instrument.

They tried the usual things. Turning it off and on again. Making sure the cables were plugged into the right spot. Replacing said cables. Testing for interference.

They tried the unusual things. Calling up the nearby army base to ask—politely—if they were transmitting at these frequencies. Cleaning the pigeon poop off the antenna. Just shooting all the dang pigeons.

I'm sure it was the oddest thing they'd ever seen. No matter where they pointed the antenna, no matter the time of day, no matter the season, there was this constant background static.

After years of banging their heads against the wall, they wondered if this hiss might be real—and might be extraterrestrial. So they sent out some feelers to the astronomical community, and before long the Dicke crew caught wind of it. They met with Penzias and Wilson. They chatted. They both came to the same conclusion: they found it. The cosmic microwave background. The afterglow of the big bang itself.

The result was two side-by-side papers, a flash heard around the world. One paper, written by the physicists, summarized the current state of the art in big bang thinking, this astounding insight that our universe was fundamentally different in the past than it is today, and that this difference is detectable and measurable.[4] The other, written by the engineers, summarized the observations.[5]

Penzias and Wilson won a Nobel Prize for their work in failing to find a source of static hiss in their fancy antenna. There's a good chance you've probably never heard of the others before you read their names a few paragraphs ago.

C'est la vie.

The cosmic microwave background, or CMB for those in a hurry, was the nail in the coffin for Hoyle's steady-state theory. The perfect cosmological principle, as lovely as it sounds, doesn't appear to apply to our universe. While Hoyle would continue to fight the good fight past 1965, the games he would have to play to reconcile the steady-state model with the overwhelming abundance of data were stretching far too thin.

Steady state predicted that the universe ought to be the same in the past. But here we are, bathed in relic radiation generated billions of years ago. It was found across the sky, almost perfectly matching a blackbody spectrum (indeed, it's the most perfect blackbody found in nature, besting even humanmade ones), with almost no variation from point to point. It couldn't be generated by stars or galaxies—their distribution is far too lumpy to explain the smoothness of the background signal. It truly appeared to be a *background*, a source of light sitting behind everything else we can see.

If you could put on microwave goggles, you could detect this bath of radiation—if only faintly. Although the CMB is the largest reservoir of photons in the universe, our universe is very, very large nowadays. But build a simple microwave receiver, and you'll pick it up. If you've ever encountered an old rabbit-ears TV that's stuck between channels, you've seen with your own eyes this fossil from a distant age—about 25 percent of the static in our lives comes from the cosmic microwave background.

The cosmic microwave background pretty much killed off any other competing theory as well. Nothing else could fit, no other idea could make the cut. The raw observational data from the past three decades were simply too overwhelming.

Our universe was different in the past, and it will be different in the future. This is the ultimate, if initially unpalatable, answer to Olbers' paradox. Why aren't we surrounded by an infinity of stars covering every square degree of the sky above? Because at a certain time in the past, there were no stars. Our universe may be old, but it's only *so* old.

The cosmos may be infinite in size (we'll get to that later), but it certainly isn't infinite in time, and it's growing larger every day. In the past it was smaller, hotter, and denser. How did it arise? What were the earliest moments like? These are very difficult questions to answer, if indeed they are valid questions at all.

But they are questions we'll eventually have to face, because the facts of our observations push us to that inevitable, uncomfortable conclusion.

Have your own personal theory of the history of the universe? That's fine—science thrives on creativity. But the knife of observation is sharp and is perfectly willing to cut your precious idea down to size. If you want your cosmology to work, you have to explain the existence and properties of the cosmic microwave background. Its presence is inarguable and its implications unavoidable.

While the subject was difficult to think about, at least theory and observations were in accord . . . for a brief moment. The angst of the nineteenth century was slowly dissolving as new understanding poured forth from the chalkboards—and now computers!—of theorists and the instruments of observers. To be clear, nobody really *enjoyed* the answers they were getting, but at least the picture of the universe was clicking into place.

The consternation about the complex nature of the stars was still there, and I'll resolve that tension in later chapters. It was replaced instead by a growing despair in the true scale, both in time and space, of the universe.

In some ways, a finite age to the universe is more troubling than the alternative. In an infinite (either in the static or steady sense), at least you can take comfort in the fact that this is just the way things are—that the universe simply persists, unchanging, through the deepness of time. But with the big bang, we know now that the universe has a past . . . and a future. And that both of these are different from the present.

While the picture of the cosmos was starting to sharpen into an unpleasant focus, at least things were making sense. Gravity, the feeblest of forces, was able to shape and govern the majesty of the heavens—Newton would have never guessed the magnitude of his initial insight! Over time, the discovery of the nuclear forces would help us understand the earliest epochs in our cosmic history, and also the mysterious processes in the hearts of stars.

The universe was revealing itself to be larger, more complex, and made of deeper stuff than we ever realized before. And while it looked, for a brief moment, like we had solved some of the largest riddles of our age—the true scope of the cosmos—seeds of mystery were already beginning to grow.

CHAPTER 7

REAPING THE QUANTUM WHIRLWIND

T he story of the initial moments in our universe, from the deepest mysteries of the first second to the exotic yet understandable plasma physics of the generation of the cosmic microwave background, has been a study in contrasts. Of radiation versus matter. Of the ungluing of forces and incredible expansion. Of detailed particle interactions leading to an imbalance of matter.

And now we've reached a point where to paint a better picture of those first instants, and to give context for what's to come, we have to turn our focus inward. Deeply inward, into the subatomic realm. When Kepler asserted that the motions of the heavens governed our lives here on Earth, and then Newton realized that the physics of gravity is universal, I doubt they would have suspected that we were going to take things this far.

For here we are, in both the story of the universe and the story of our understanding of it, at a point where our knowledge of fundamental physics doesn't just govern arcane and complicated interactions in particle colliders. No, it determines the history and even fate of the universe at the largest scales.

Over the centuries we've come to realize that it's not just the laws of gravity that hold across the heavens and the Earth. The same goes for every force, every law, every interaction. Thermodynamics, electromagnetism, nuclear physics, the whole lot are what bind us to the cosmos. We may not fully understand the initial moments of our universe, but we are not afraid of attempting an explanation. We cannot visit the era of recombination and the birth of the cosmic microwave background, but we can recreate it—in miniature—in our laboratories. The inflationary epoch is inaccessible to direct observation, but we can probe it with mathematics.

The universe across both time and space is hopelessly messy, but in a good and bad way. Bad because it makes it *much* harder to understand than we previously thought. But good because it's just as messy as our experiences here on Earth—which means we can perform experiments, test ideas, and form hypotheses to guide us. Sciencey stuff.

Is physics truly universal? Do the laws and relationships we reveal in this place at this time hold across the cosmos? I'll get to that question in a later chapter, but for now we can rest assured that it seems on all accounts to work. And the perfect starting place is the humble spectral line.

Max Planck wasn't directly working on the problem of spectral lines, but his simple but pioneering work laid the pavement for the road to quantum mechanics—which *does* explain spectral lines.

Max was working on the blackbody problem. Remember all that stuff about blackbodies? Of course you don't—time to reread the last chapter. The hotter the thing, the brighter and bluer it glows, and the cooler the thing, the dimmer and redder it glows. What's the beef? The deal was that while all those relationships were sorted out experimentally in the late 1800s, nobody could *explain* it. You know, with physics and math.

One of the best models we could come up with, thanks to physicists Lord (John) Rayleigh and Sir (James) Jeans, was pretty straightforward: the atoms and molecules in a blackbody dumped some of their vibrational energy into radiation, which will get emitted, making it glow. But as far as their physics could tell, the transfer of vibration to radiation energy was totally egalitarian: some energy would go to low-frequency radiation and some to high-frequency radiation.

Given the pedigree of the originators, it's a surprisingly communistic approach to physics: from each frequency according to its ability, to each frequency according to its need. While this approach works in a limited set of cases, it quickly broke down in the wonderfully named "ultraviolet catastrophe." If all frequencies each get a little bit of energy, then any common household object ought to be emitting everything possible, including high-energy ultraviolet rays, X-rays, and even gamma rays!

This, uh, doesn't happen, which everybody realized but nobody could figure out why.

It took regular guy Max Planck to come up with a solution: you gotta pay to play in this game. In his attempts to coerce the mathematics to fit the observations—to provide a halfway decent explanation for the blackbody phenomenon—he introduced what he considered to be an ugly hack: *quantization*. If he assumed that radiation couldn't be emitted with any energy level it pleased—if radiation came in discrete packets—then his equations worked.[1]

Those packets are called *quanta*, which means, well, packets. Compare a glass of water to a bag of potato chips. I know that water is made up of zillions of tiny molecules, but at the human level, it's a continuous fluid: you can have any amount of water you want, from the teensiest drop to the gushiest geyser. But your potato chips are *quantized*. You can, if you're hungry enough, have a lot of potato chips. But you can't have less than one—a single potato chip is the quantum limit of the bag. And your choices for the number of chips are always whole numbers: one chip, two chips, twenty-seven chips (slow down there, fella), and so on.

Yes, I know in reality that you can break a chip in half, smarty-pants. But just roll with the analogy; it's the best I could come up with, probably because I'm hungry.

So Planck fudged the math to make radiation behave less like water and more like potato chips, and this solved the ultraviolet catastrophe. To make one "chip" of radiation (let's call it a "photon"), you need to expend a fixed amount of energy. For a given temperature, lower-frequency radiation is easy to make: each individual photon takes just a tiny amount of energy to manufacture, so you can spit out a lot of them.

But the high-frequency photons take a lot of energy just to make a single one, and if you only have half the required energy, or three-fourths, or 99.999999 percent, it's not gonna happen. You have to grab the whole chip or you don't get any chips at all. This explains why we're not awash in cancer-inducing radiation from a hot cup of coffee or cookies fresh from the oven: they don't have enough energy to produce the hard stuff.

This may sound obvious now, after the world has had a hundred years to get used to the idea, but back then it was pretty radical stuff. Even Planck himself didn't really take it seriously: he was willing to try anything to get the mathematics to work, even this, but he considered it a stopgap measure until something better came along.

Nothing better ever did come along, but at least he ended up with a Nobel Prize for it.

That fundamental relationship between the frequency of a photon and its energy birthed a new constant of nature, one that told us about the ground-state potato-chippiness of reality: Planck's constant, which we first met way back in the earliest, sketchiest moments of the universe. It's just a simple number with no cool superhero origin story. Planck himself calculated the necessary ratio using all the known blackbody experimental results. It was a kludge, an ugly hack, a number tossed in to make the math work.

And just a few years after his initial preposterous proposition, Einstein continued the game by studying the so-called photoelectric effect, positing that it's not just the *emission* of radiation that's quantized (which is all you technically need to explain the blackbody effect) but its absorption and transmission as well.[2] Radiation of all forms only comes in discrete little packets.

And then physicists went *nuts*. What if it wasn't just light that was quantized, but, like, everything? What if all energy was quantized? What if—bear with me here—our fundamental reality is just a bag of potato chips?

Like I've said before, in science you're free to say whatever crazy thing pops into your head, but if you want to play the physics game, you have to think through the consequences of that crazy idea and test those consequences against observations.

One consequence is the nature of the atom, a subject under considerable debate and study in the opening decades of the twentieth century. The same time that astronomers were pushing the boundaries of the extent of the cosmos, physicists were trying to probe the tiniest structures known. Relatively quickly it was realized that an atom is composed of a small, dense, positively charged nucleus (a bundle of protons and neutrons), surrounded by a buzz of distinct negatively charged particles (the electrons).

The electrons were bound to the nucleus but could be knocked off if given a sufficient kick. Additional electrons could be added to an atom, which would change some of its chemical properties but otherwise leave it the same.

OK, fair enough, but the major question was how electrons arrange themselves in an atom. If you just consider them as little electrically charged balls whizzing around a nucleus like planets around the sun—which has inexplicably become the universal default symbol for "Science!"—it just doesn't work. Electrically charged balls whizzing around emit radiation, which saps energy, which should send them crashing into the nucleus. They don't, so they aren't.

The answer is potato chips. Electrons, bound to an atomic nucleus, don't get to have any sort of energy they want. No, there's a minimum energy level that they can settle into—their behavior around the nucleus is quantized. This prevents the electron from slamming into the nucleus. It simply can't, because the quantum nature of reality prevents it from having a fraction of its minimum energy. You can only have one chip, not half a chip, and an electron can only get so close to a nucleus, and no closer.

(I need to add that there's a lot more math that goes into it, including the wave-particle duality that is another quantum facet of nature, but I think this is enough to get my point across.)

And just like for photons and radiation, the energy levels of electrons around a nucleus come in discrete steps. An electron can be on step 1 or step 14, just like you can eat chip 1 or chip 14, but it can't be between steps. So when you add or remove energy from that electron (say, by having it absorb or emit some radiation), it can only do so in discrete chunks.

In other words, the radiation that an atom or molecule will absorb or emit will only correspond to specific energies, which for radiation means that the light going in or coming out can only be specific wavelengths. So a hot gas of one particular element will emit very distinctive light, not everything in the rainbow. A distinct pattern of lines in the spectrum. If that same gas is blocking a light behind it, it won't absorb all of it, just a few frequencies here and there.

So the curious but incredibly useful phenomenon of spectral lines that astronomers had known about for decades, used to unlock the vast expanses of the cosmos from the motions of distant stars to the expansion of the universe itself, owes its unique properties to the subatomic quantum nature of energy.

Thanks, Max!

Once the quantum revolution got seriously underway in the early twentieth century, physicists really went to town exploring the weird and wonderful world lying in between our molecules. At first blush, it's an odd coincidence that at the same time physicists were busy unlocking quantum mysteries, astronomers were revealing cosmological profundity, but really this was standard operating procedure. Different tribes of scientists will explore different realms of nature, carving out little investigative niches for themselves, Sherlocking the clues wherever they may lead.

Scientists don't realize the full implications of their work, both for their own fields (often) and for disciplines outside their own (almost always). As we saw above, the chemists of the nineteenth century would have never realized that their investigations of hot gases would unlock the motions of galaxies, or that Max Planck's awkward studies of blackbody radiation would end up being applied to relic radiation from a distant epoch of our universe.

The same goes for Paul Adrien Maurice Dirac, whose fabulous antimatter hotel I introduced earlier and who has such a fantastic full name that I just had to type it again (plus he authored his papers as "P. A. M. Dirac," so I'm guessing he wanted us to remember the whole thing). Dirac was practically diabolical in his studies of theoretical physics—he had a useful/nasty habit of simply making up from whole cloth mathematical functions and operations just to push his understanding of physical relationships.[3]

It seems he often went on hunches, and if a new formalism worked, it worked, and he left it at that. Later generations of mathematicians, much to their consternation, were frequently forced to go back over Dirac's work and provide the necessary proofs to validate his insights.

But enough on the awesomeness of Dirac, one of my favorite physicists of all time (the other is James Clerk Maxwell, whom we already met, if only briefly).

Dirac was interested in a lot of the same problems that other now-renowned physicists, including Planck, Einstein, Schrödinger, Heisenberg, and Pauli (and tons of others), were in the 1920s: the strange mechanics of sub-atomic particles. One by one, experiment by experiment, scientists were beginning to piece together the rules that particles lived by.

One of those rules was a newfound property called *spin*. If you shoot some neutral atoms through a magnetic field, those particles won't even notice, and they'll just sail on through unmolested—that's one of the benefits of being "neutral." But in 1922, Otto Stern and Walther Gerlach (as a side note, you may notice a lot of German names associated with quantum mechanics; they were kind of dominating theoretical physics at the time) noticed something fishy. When they sent totally neutral silver atoms, with an equal complement of positively charged protons and negatively charged electrons, through an inhomogeneous magnetic field (that just means that the "north" direction in the apparatus was slightly stronger than the "south"), the atoms would get deflected.

Hmmm. Neutral atoms, affected by a magnetic field. The only way to explain this result was if the fundamental particles buried inside the atom had a new property, something like the usual mass and charge that allowed them to respond to magnetic fields. By analogy, the physicists turned to spinning metal balls—if you charge up a metal ball and set it spinning, it acts like a magnet, and if you throw a magnet through an inhomogeneous magnetic field, it gets deflected.

Spinning charged metal balls it was, and so the new property was dubbed "spin." Let's ignore the fact that particles like electrons are modeled to be infinitesimally small and aren't really, you know, metal balls and don't really, you know, spin. Think of it more as "magnetic response" and you'll save yourself some sanity.

It gets even weirder. The property of "spin" possessed by our subatomic brethren itself is quantized in an odd way: for an electron, for example, it only comes in two varieties, dubbed "up" and "down," probably because the deflected silver atoms shot through Stern and Gerlach's experiment split up into two distinct clumps, one on the upside and one on the downside.

Through a ridiculous number of follow-up experiments, the scientific community devised a set of "rules" to explain what nature was up to . . . down there. These rules of spin explained all the results (ahem, because they were designed to) but didn't explain where this property came from. The mystery of the behavior of the supposedly neutral silver atoms was "solved" by the newly devised rules of spin: electrons like to pair up spinwise up-to-down, canceling out their effects, but with silver, there's a loner odd-number electron hanging out. It's the unpaired spin of this single electron that's responsible for the experimental results.

Like I said, the early quantum mechanics couldn't explain this spin effect naturally; it didn't pop out of the equations. At least, until Dirac took a crack at it. He wasn't directly trying to solve the spin problem (notice a pattern?), but a more pernicious one: reconciling the burgeoning rules of the quantum world with the lessons from Einstein about special relativity, which by this point was almost a couple dozen years old and generally accepted to be really, really important.

Special relativity itself is almost like a metatheory of physics: the fundamental relationships between space and time and mass and energy must be integrated into *any* theory of physics for it to be considered fully correct and universally applicable. As we saw earlier, Erwin Schrödinger attempted a reconciliation of special relativity and the mathematics behind quantum mechanics and gave up; he couldn't make heads or tails of what the chimera would mean.

So he decided on a simpler approach, declaring the incorporation of relativity to be Somebody Else's Problem, and produced a not-fully-correct-but-still-really-useful equation that students still today learn in Quantum 101.

Dirac tackled the same spin problem and initially also couldn't make heads or tails of the mathematics, but he just forged ahead anyway because he was Dirac and Schrödinger wasn't. By pretty much inventing a formalism from whole cloth, his solution to the quantum dilemma naturally incorporated the concept of spin—it popped right out of the equations. If Dirac had come around to this problem a decade earlier, he would have *predicted* this new quantum mechanic property of fundamental particles.

There was another consequence of Dirac's solution. Or rather, solutions. He identified a symmetry in the equations that predicted that every particle had an evil twin: one with the same mass, spin, and all the other properties, but with a perfectly opposite charge.

Antimatter.

The ensuring decades saw a gold rush of physicists hunting for new particles—and striking gold. Unfortunately this isn't a book on fundamental physics, except where it intersects our story of the universe (please call Prometheus Books and beg for *Your Place in the Quantum: Understanding Our Tiny, Messy Existence*), but since the early moments of the big bang were such a hot and crazy time, and the deep future of our universe will see a resurgence of some strange characters, we need a nanozoology lesson.

At the highest level, you have two general kinds of particles: the fermions and the bosons. Extremely generally, the fermions are the things, and the bosons are carriers of the forces. If I chopped up your molecules into atoms and cracked open the atoms, I would find a bunch of fermions: protons, neutrons, and electrons. To do the cracking I might employ a giant laser, which is made up of photons, the boson carrier of the electromagnetic force.

Hence the name *W boson* for the carrier of the Weak force. The pickup trucks of the strong nuclear force, the gluons, are also bosons and do exactly what their name suggests (if you recall). Gravity doesn't formally have a particle-based carrier known to current theories, but that boson is called the graviton anyway.

The only difference between the fermion camp and the boson gang—besides whom they're named after, Enrico Fermi and Satyendra Nath Bose—is their spin. The way the conventions in the mathematics worked out, fermions are defined to have "half-integer spin" (like $1/2$, $3/2$, $5/2$, etc.) and the bosons have "whole-integer spin" (like 1, 2, 3, etc.). That may seem like splitting hairs, but for reasons I won't go into, this difference has profound implications for how the particles mix together—all the fundamental differences between how particles like photons behave and how particles like electrons behave can be attributed to their different spins.

And it turns out that protons and neutrons aren't fundamental; they're made of smaller dudes named quarks. There are six quarks: top, bottom, up, down, strange, and charm. Don't ask about the names.[4] Quarks normally either pair up (and the buddy-system pair of quarks is now dubbed a "meson") or run around as triplets (now called baryons).

Baryons are far, far more common than mesons, hence why we focused on them so much in the story of baryogenesis in the early universe and their ultimate victory over their evil, mustachioed antimatter twins.

Just in case you were wondering if the electron was feeling lonely, don't worry. There's a version of the electron that has the same charge and spin but is about two hundred times more massive; it's called the muon. And a heavier version still is called the tau. Paired with each of these is a related neutrino. Neutrinos themselves are nearly massless, ghostly particles that hardly ever interact with anything, ever. So there's an electron neutrino, muon neutrino, and tau neutrino.

The electron, its two heavier siblings, and the only-visit-for-Christmas neutrinos form a group called the leptons.

So there are six quarks and six leptons. The quarks feel the strong force, so they bind up tightly into balls that we call protons, neutrons, kaons, pions, and so-ons. The protons and neutrons are especially friendly and bunch up into atomic nuclei. While the strong force is, well, strong, its limited range confines its influence.

The leptons don't feel the strong force, so they just hang out by themselves.

The weak force transforms one kind of quark into another and, in the process, emits one of the electron triplets along with its associated neutrino. As we saw earlier it's a pretty intense force when conditions are right, which they hardly ever are in the present-day universe. It too has a rather short range, hence why it was the last force to be discovered.

The electromagnetic force will touch anything it can, and given that the photons are massless, it has infinite range. Think about it: that star you see twinkling in the night sky hurled its light thousands of years ago, and across the vastness of empty space, it was able to activate the rods and cones in the back end of your eyeball. That's a pretty impressive feat for a force, but electromagnetism does have one weakness: it can only care about you if you have an electric charge. If you're neutral, you're invisible to photons.

The last known force, gravity, is so weak it's hardly worth mentioning; and indeed, when it comes to particle physics it doesn't even enter into the calculations. As far as the inner workings of an atom are concerned, gravity might as well not even exist. But as we saw in the early years of the universe, it does have two superpowers. One: like the photon, it has infinite range. The gravity of that distant star isn't affecting you nearly as much as Kepler would have preferred, and at the barest fraction of what the photon can do, but technically it does have *some* influence.

This very book, sitting right in front of your face, has a greater gravitational influence on you than even the largest planets in our solar system. So there.

The second superpower is that gravity acts on *everything*. If you have energy or mass, gravity is going to touch you. So even though it's weak, it's persistent, and in the cosmological game you definitely get an A for effort.

Did I mention the antiparticles? Every lepton—every electron, muon, tau, and neutrino—and every quark—all six of them—has an associated antiparticle.

It's a zoo run by an insane zookeeper, so don't feel inadequate if it seems like it's just a torrent of names, jargon, and stats. I barely even scratched the surface of particle interactions, but I wanted to at least introduce them because it's about to get a lot funkier in here.

Following Dirac's lead, multiple scientists continued exploring the connection between relativity and quantum mechanics. The play of the game was simple—take each of the four forces in turn, shove special relativity up its rear end like a Thanksgiving turkey, and hope for the best.

First up to bat was electromagnetism (it was an obvious choice, given that it was this very theory that prompted Einstein to develop relativity in the first place). Unfortunately, the best was looking pretty peculiar. To make everything make sense, theorists had to revisit the concept of the *field*.

A field in physics is a lot like a cornfield, except mathematical. As you walk around the cornfield, you'll notice that the stalks are different heights. You could make a map, if you wanted, that assigned a particular cornstalk height to each position on the ground. That way, when you reported back to the homestead, if someone asked you about the height of corn at a given latitude and longitude, you could smartly and promptly return an answer.

Congratulations, you've made, uh, a field. A mathematical one. It's a set of values assigned to coordinates in space. The number of dimensions doesn't matter, and for our purposes we'll jump ahead from the farmer's back forty to the full four-dimensional space-time that we're now accustomed to.

Fields are used all over physics; you may have even heard of them used in casual conversation, depending on your definition of "casual." Take, for example, the electric field. If you put a single isolated positive charge in the middle of space, we can, for the sake of mathematical convenience, assign a *field* to that electric charge. That field will tell you, wherever you may be located in respect to the isolated charge, how you might respond to it. If you yourself are positively charged and close by, the field will instruct you to be suitably repulsed. If you're negatively charged and far away, the field will whisper sweetly in your ear to come a little closer.

Electric fields, magnetic fields. Unified together into the electromagnetic field. The gravitational field. By the twentieth century, fields were about as common as . . . fields. They were already seen to be more than a mathematical convenience, and after the work of Dirac and others, they took on an entirely new property.

The fields associated with our normal everyday experience, like the electromagnetic field, aren't just passive conveyers of information—they are *alive*. They are dynamic. They flex and bend and wave. Waves of electricity and magnetism are already familiar to us: that's simply light. Good old radiation.

Here's the juicy bit. When quantum mechanics and special relativity unite to take a stab at fully explaining the electromagnetic force, it becomes *all* about the waves. In this picture the entire universe, all of space-time, is filled with a field, the electromagnetic field. Ripples can propagate through this field, which we correctly identify as light.

But it is at its heart a quantum field. And the number one rule of the quantum world is that nothing is smooth; everything comes in packets, or chunks, or bits, or units: quanta. Thus waves of the electromagnetic field can only be so small for a given amount of energy; there's a minimum wave size on the field: the photon. The familiar particle version of electromagnetic radiation.

Now take this concept that doesn't sound so radical for radiation and extend it. To everything. Look down at your hand. You know that if you could examine it closely enough, you would see tissues composed of cells composed of molecules composed of atoms composed of subatomic particles. Those particles, the electrons and quarks buzzing around, are simply vibrations: excitations of a field that permeates all of space-time. This is the world of quantum field theory, the crowning achievement of twentieth-century physics. In this picture the field is the ultimate physical object. There is one field associated with each type of particle, from the electromagnetic field for radiation to the suitably named Dirac field for electrons.

The fields fill up the universe and overlap each other, constantly awash with waves and counterwaves and excitations and propagations. A ripple rushing through one of these fields in space-time could represent anything: a cluster of protons racing from a supernova explosion, a brief weak force interaction, or the dance between gluons and quarks inside a nucleus.

These fields can support localized vibrations, but since they are distinct physical entities in their own right—arguably more "real" than the particles we associate with them—they can change and evolve under their own rules. Thus creatures like the hypothetical "inflaton" field can dynamically change, and since that field permeates all of space-time, it can radically impact the rate of expansion in the earliest moments of the universe—inflation.

It's admittedly hard to think about particles as anything but particles. But some of the peculiar properties of particles click into place once you adopt an appropriately quantum field worldview. Like the fact that they can be created and destroyed at will.

Einstein taught us mere mortals that energy is mass. Feel free to ignore the c in the famous equation $E = mc^2$, since almost all theoretical physicists do anyway. It's just a constant, a number, floating around in the math. Here it serves as a conversion factor; it tells you how much energetic punch a given amount of matter can pack. When you remove that pesky term, Einstein's relation becomes a lot clearer—and a lot more awe-inspiring: Energy is mass. Mass is energy. They are equivalent. They are the same.

Once this fact settles into your bones the world becomes a lot stranger. You can disappear a particle without any magic tricks: if it has mass, it can convert to energy. And if you bunch enough energy together at a specific place, boom—you've made yourself a particle. We're actually much more used to this process than we think: every time you turn on a light, trillions upon trillions of photons—the particles that carry the electromagnetic force—are spontaneously born at the light source, leap across the room, and are immediately extinguished in a deposit of energy at your eye.

For photons, this creation/destruction process is pretty easy. They're massless, after all, and so don't take a lot of energy to get up and going. But if quantum field theory has taught us one thing, it's that what goes for electromagnetism goes for everybody. Thus electrons, quarks, W bosons, whatever-ons, can all be created in a flash and snuffed out just as easily as flicking a switch.

It's hard to visualize unless you have on your field glasses. But if particles are just vibrations in a field that stretches across all space-time, then *of course* they can be created or destroyed at will. It's just a matter of adding or subtracting vibrational energy to or from that field. Pluck a guitar string, make a note. Pluck a quantum field, make a particle.

The collisions and battles of the early universe take on a new light in this context. Just as a single field can populate a volume of space with a bunch of particles, the various fields can interact with each other—a vibration in one can translate to a vibration in another. This is how we understand particle transformations, how a photon can split into a particle-antiparticle pair or vice versa. The struggle of matter versus antimatter in the early universe was more about the relationship of fundamental quantum fields, vibrating against each other for dominance.

And just as the rules of quantum mechanics dictated a minimum energy level for electrons bound to an atomic nucleus, these fields that flit and float throughout the universe also get stuck with a minimum energy. A certain low-level buzzing, a humming that permeates the cosmos. This is usually described as the "quantum foam" and visualized as "virtual" particles constantly popping into and out of existence before you have a chance to notice. That's not an incorrect description, but (a) I'm not going to bother untangling the word "virtual" for you, and (b) I like the concept of a background hum better. It's more Keplerian.

This ground-state energy of the universe has a more formal name—vacuum energy. It's a real thing, and I'll turn to it in more filling yet ultimately unsatisfactory detail later.

The complete picture of our universe at the fundamental level, from the catalogs of particles to the wiggly-field nature of interactions, goes by the astoundingly boring name of "the standard model." This (complicated) model categorizes the leptons and baryons, puts antimatter in its place, incorporates the Higgs boson, and provides a fully quantum picture of electromagnetism and the weak and strong nuclear forces.

As standard as it claims to be, this massive monument of modern physics isn't as complete as we'd like it to be. We know of some creatures in our universe that don't fit under our one-size-fits-all umbrella, and they'll get some attention in our discussion very soon.

Oh, and there's no explanation for gravity. At the current state of play in our understanding of the universe, gravity stands alone. The force that Newton identified as universal, the force that swung the planets in their orbits so regularly that Kepler could find a deep pattern in their motions—the force that, perhaps, we're most intimately familiar with—does not have a home in our quantum-backed view of the world.

As beautiful and complex and elegant as general relativity is, we know it's incomplete. It never got the quantum makeover that the other forces did. All attempts to unify it under a comprehensive picture (a "theory of everything") or, failing that summit, to settle for a "just a theory of quantum gravity" base camp, have utterly failed.

It's a problem of infinities. Quantum field theory calculations are notoriously difficult; the mathematics have a tendency to blow up in unexpected and unwanted places, rendering further calculations (and any predictions) useless. For a long time it was thought that quantum field theory would be completely unusable as a physical theory. Then, in 1948, Richard Feynman, Julian Schwinger, and Sin-Itiro Tomonaga showed a path forward, essentially by tucking away all the unpleasant infinities into a small collection of terms, then replacing those terms with known constants like the electron mass.

It was, as usual, an ugly hack, but it worked, and the entire standard model eventually flowed from these methods. But gravity remained resistant. Including a dynamic space-time in the calculations of fundamental particle interactions adds too many infinities to package up; it's ten gallons of infinities in a five-gallon bucket.

This is why the initial moments of the big bang are so fascinating. Not just because, hey, we like to know stuff about the universe, but also because this is the era when quantum gravity actually mattered. We could never find a solution for bringing gravity into the quantum fold, and doing so wouldn't affect our daily lives one bit. But if we're trying to build a complete picture of the universe, which we have been since the days of Kepler and Galileo, then we need to crack the gravitational puzzle to unlock those first moments.

Our universe was born in mystery, but that doesn't mean that puzzling specters don't still haunt the modern-day cosmos. By the 1960s we had firmly established both the big bang and beginnings of the standard model. We knew the universe both inside and out. That would perhaps be the last time we would feel so assured in our knowledge.

A GUIDE TO LIVING IN AN EXPANDING UNIVERSE

Let's just go ahead and assume that you're a little confused. Don't worry, I won't judge. We've all been there. Trust me. Answers cosmological don't come easy—it's not like our brains were hardwired for pondering the expansion of a universe billions of light-years across, so there's bound to be some poor connections made between the mathematics that we use to describe our reality and the words that I'm using to describe the mathematics. So let's take a little break from our breakneck race through cosmology and cosmological history to settle some old scores, once and for all.

Like the good teacher I pretend to be, I've tried to anticipate some of your questions:

WHAT DOES IT MEAN FOR THE UNIVERSE TO "EXPAND"?

Although the analogies come in handy in some limited cases, I want you to stop thinking of balloons inflating, bread rising in the oven, or anything else. Just simply absorb this bare-naked observational statement: on average, at large scales, all galaxies in our universe are moving away from each other. Take a snapshot of all the galaxies in the universe and record the distances between them. Wait a while. How long? It depends—how long is your ruler? Take another snapshot and repeat your distance measurements. The second set will be bigger numbers than the first set. And there it is: the expansion of the universe.

WHERE IS THE CENTER?

It's right here, where you're standing. Surprised? I thought we junked that whole Earth-centered business. The new wrinkle is that "You're the center of the universe" is now a meaningless statement. From any perspective, anywhere, on any planet in any galaxy, they will view themselves to be the center. Since every galaxy moves away from every *other* galaxy, the "center" is arbitrary. From our perspective, it looks like everybody is fleeing from the Milky Way. From Andromeda, same story. Pick the farthest, dimmest galaxy you can spot. If there are aliens in the galaxy, they'll come to the same conclusions: the universe is expanding from *them*. "Center" doesn't have any meaning in an expanding universe.

WHERE DID THE BIG BANG HAPPEN?

You really won't let this go, will you? The universe has no center. The big bang didn't happen somewhere over there. It happened everywhere. It happened in the room you're sitting in, in the distant galaxies, and in all the voids between. There's no place in the sky you can vaguely gesture to and say, "It happened roughly over there." The big bang wasn't an explosion *in* space, it was an explosion *of* space. It happened everywhere simultaneously. It was a distinct moment in time that sits in the finite past of everything in the universe, not a point in space that everyone can, er, point to.

WHERE IS THE EDGE OF THE UNIVERSE?

Just as it has no center, it has no edge. I know, I know, it's very hard to imagine something with structure but without any outside. Look, if the universe had an edge, like a wall, separating universe from not-universe, then the not-universe would be a *thing*, and since by definition the universe is *all the things*, we would have to include it in the definition, and we'd be right back to where we started. If the universe is infinite, then we don't even have to think about it anymore.

If the universe is finite, it gets a little harder to contemplate, since we're not used to the concept of objects having limited spatial extent but no boundary.

CAN I GET A METAPHOR, PLEASE?

Fine. Imagine the surface of the Earth. Just the surface, nothing else. Every location on the surface can be perfectly described by two numbers: latitude and longitude. Where is the center of the surface of the Earth? Notice I did *not* say "the center of the Earth." I'll ask again to really make it clear: where is the center of the *two-dimensional surface* of the Earth? There isn't one. Where's the edge of the surface of the Earth? There isn't one (in two dimensions). Does the surface of the Earth have finite size? Yeah, 196.9 million square miles. Does it have a boundary? No.

ARE YOU SAYING THAT THE UNIVERSE IS CURVED LIKE THE EARTH?

Yes, but no. The universe may be curved at very large scales, but it doesn't matter. The point is that finite structures don't need to have a boundary in order to be a thing. Sure, we can easily visualize the surface of the Earth by imagining it embedded in a higher three-dimensional setting, but it's a mistake to think that everything must be embedded like that. The three-dimensional universe (ahem, four-dimensional, but let's focus on just space right now) *could* be embedded in a four-dimensional spatial structure, but it doesn't *have* to be for all the math to work out. It's just one of these weird things where the math is totally chill about it, but it's hard to express in English, let alone paint a picture in our brains. We're just not wired to conjure up images of three-dimensional, expanding, finite things. Sorry.

I THOUGHT YOU SAID THE UNIVERSE WAS FLAT.

One, that's not a question, and two, you're skipping ahead in the book.

SO WHAT IS IT EXPANDING INTO?

OK, I can't be too hard on you for not cracking this nut. The answer isn't "nothing" or even "something else." The universe has no outside, and no edge. This is the real kicker: this question doesn't even make sense when it comes to the universe. There simply isn't an answer, but not because we don't know, but because the question can't be formulated. "What color is a cat's meow?" has all the correct English words and grammatical structure, but it's not a question we can answer. There is simply *no such thing* as an outside to the universe.

HOW BIG IS THE UNIVERSE?

Current estimates (which are very good, thank you very much) pin the whole observable universe to be a sphere about ninety-three billion light-years across. That's a big universe.

HOLD ON A SEC, WHAT DO YOU MEAN BY "UNIVERSE"?

I'll admit it, I've been playing a little fast and loose with my definitions. In one chapter I say that the universe is impossibly large—due to inflation—or maybe even infinitely big, but here I just gave it a size. When people (like me) refer to the size of the universe, we usually mean the *observable* universe, the limit of what we can see based on the age of the universe. It's the distance to the "horizon," which, like a horizon on Earth, is . . . the limit of what you can see. The cosmic microwave background sits juuuust inside this horizon, and so for all practical purposes, it is the farthest thing we can see. But there is much more universe that we can't see—more galaxies, stars, solar systems, farm fields. That's the whole entire universe, but since it's completely inaccessible to us, can never affect us, and frankly doesn't care much for us, it doesn't matter. At all, so just ignore it.

UH, OK. WELL, HOW OLD IS THE UNIVERSE?

13,799,000,000 years, plus or minus 21 million years "since" the big bang. Yes, we know it that precisely,[1] and yes, we're very proud of ourselves.

DOESN'T THAT MEAN THE UNIVERSE EXPANDS FASTER THAN LIGHT?

Well, yeah. The radius of the universe is bigger than 13.8 billion light-years. So *superluminal* expansion is a thing, especially during inflation (that was kind of the point). And even now! In an expanding universe, the farther you go from the Milky Way, the faster the rate of recession. Double the distance, double the speed. Eventually you'll come upon a distance that gives a speed faster than light. What's the big deal? Oh, you think that's a problem with special relativity, which set a universal speed limit? Haven't we, like, tested that or something?

Come here, child; let me share a secret with you. Special relativity is a *local* law of physics. The quantity we call "speed" is really only something you can measure nearby: you'll never measure the speed of a rocket blasting in front of your face to be greater than light. But a galaxy on the distant edge of the universe? It can have any "speed" it wants—if its redshift implies that it's going faster than light, it just means that the expansion of the universe is carrying it away so fast that you'll never, ever catch up to it. Even if you really want to.

HOW CAN WE ALL AGREE ON THE AGE? ISN'T TIME RELATIVE?

Yes, time is relative and not all clocks tick at the same tock . . . in *special* relativity. But the game of cosmology is played with *general* relativity, which is, you guessed it, more general. While usually time and space are all mixed together in a single cotton-poly blend of a fabric, in a homogeneous, isotropic universe (like the one we live in), time splits off from the spatial dimensions and ticks away comfortably at its own preferred rate. Due to the way the cosmic microwave background was emitted (across the entire universe at about the same time in about the same way), you can use that to find the heart of the cosmological clock. No matter your motion through the universe, once you measure the CMB, you can compute a frame of reference that is, at rest, relative to the universe. Once you know that, you can measure the universally common time (called the *conformal time* if you're feeling fancy) since the big bang.

IS MORE SPACE BEING CREATED, OR DOES THE EXISTING SPACE STRETCH?

Does it matter?

HOW CAN GRAVITY MAKE THINGS EXPAND?

It's not that gravity is, per se, making everything expand. It's that the behavior of the universe writ large is determined by its initial conditions and its contents. In the framework of general relativity, the universe found itself in an expanding state, and so that's what it's going to do unless somebody tells it otherwise.

IF THE EARLY UNIVERSE WAS SO DENSE, WHY DIDN'T IT COLLAPSE INTO A BLACK HOLE?

Remember that black holes, those infinitely dense bastards of the cosmos, are things *in* space. It's hard for *all of space* (i.e., the universe) to be a black hole: it's a different setup. The catastrophic, unstoppable gravitational collapse that leads to the formation of a black hole is driven by differences in density from place to place. But the extremely early universe was equally extremely uniform: despite its overall global density, there wasn't much difference within the cosmos at the time. It's only later that contrast arises.

IF BOTH SOLUTIONS ARE POSSIBLE, WHY IS THE UNIVERSE EXPANDING RATHER THAN CONTRACTING?

Now *that's* a good question.

BEHOLD THE COSMIC DAWN

Have you ever played one of those games where you look at a super-zoomed-in image of a common household object and try to guess what it is? It's unreal. Below the threshold of normal human vision, everything just looks so freaky. What appears to be alien landscapes or monsters from your worst nightmares turns out to be shoelaces or stains on a coffee mug. When the micro is brought into the macro, when structures usually hidden beneath detection come into the realm of normal experience, it's unsettling.

A few hundred thousand years in, our universe needed a little unsettling. After the release of what would become the cosmic microwave background, the light was free; the universe quickly became dark. With the inevitable cosmic expansion that dictates so much of the physics of this story, the light that once so brilliantly flooded the universe shifted down from the intense white-hot energy of its initial release to a smoldering red, eventually slipping out of the visible altogether. For the first time in its exotic, strange journey, the universe experienced true coldness.

As the universe celebrated its one-millionth birthday, it was practically geriatric. Nothing substantial or worthy of note had happened to it since the flash of recombination. No radical phase transitions. No battles between competing forces. Matter had finally dominated over all other forms of energy density—it was now the big cheese in the universe, but it inherited a bleak, featureless landscape.

It seemed as though the universe, before it even had a chance to really get going, was locked into a grim, boring fate. Like when you're stuck watching a movie you don't really care for, it resigned itself to sit back and maybe at least catch a nap. Cooling and expansion—that's it. That's all that is left for the universe to do. You can't fight gravity, after all, and it's the dictates of general relativity that put the universe in this state.

The expansion is even slowing down—unbelievable! With matter in charge, a slowdown is inevitable. The expansion rate of the universe is intimately wedded to the proportion of the various contents within it, and with radiation not just losing dominance but rapidly plunging to essentially insignificant levels of background annoyance, the expansion has to listen to the voice of matter and matter alone. And that matter is self-gravitating: it pulls on itself and everything around it.

You make a planet, and the planet will want to collect an atmosphere and some satellites. You make a galaxy, and the galaxy will attract the flows of gas surrounding it. You make a universe full of mass, and that mass will *pull*, resisting the primordial urge to continue expanding and separating away from itself. The universe may have been born in a state of expansion, but matter is the stereotypical stubborn donkey, refusing to budge. Depending on the amount of mass, the expansion may just gently slow down over the eons, or it may give up entirely, stopping the expansion and pulling the universe in on itself.

Whatever the long-term fate of the universe, at this stage a million years into its history, the density of matter was high enough to begin slowing down the expansion, which was probably a new and uncomfortable sensation for the cosmos. That's the ultimate picture of the universe after matter's triumph—in the postrecombination cosmic landscape, we live in an expanding (less vigorously, but still pretty spry for a creature a million years old) universe, flooded with a reddening population of photons and an abundance of newly forged hydrogen and helium atoms, which simply floated around dumbly.

About that structure. As we saw when astronomers and physicists first started cooking up our modern view of cosmology, they adopted what would become known as the cosmological principle: our universe is pretty much the same from place to place. What started as a handy trick for simplifying the crazy-hard mathematics of general relativity sort of evolved into a standing rule, taking a cue from Copernicus and turning our non-specialness into a generic facet of our cosmos. The universe is large and generally uncaring about us and really quite boring.

This principle of cosmological insignificance—perhaps the greatest differentiator between the cosmologies of our ancestors and our modern view—was validated in a big way by the detection of the cosmic microwave background. As predicted, it was pretty much unchanged throughout the sky. And 41,253 square degrees of the same dang thing is a pretty clear signal that even the most clueless could pick up on. The evidence validated the core assumptions of the big bang model.

But at some scale it breaks down. Sure, if you compare different parts of the universe at big scales, those parts look statistically the same. But below those scales you have galaxies, stars, planets, people, pumpkin spice. And those are most certainly *not* the same everywhere you go. The universe isn't wall-to-wall stars; there are gaps. The Milky Way looks different from Andromeda, and way different from the Magellanic Clouds.

There's structure. There are differences. There are regions of incredible density and places with only a few scant hydrogen atoms. There are stars and not-stars. The universe is not a completely uniform tomato soup—it's a chunky beef stew.

How can we resolve this apparent paradox? To fit the observations (the raw data that all our ideas are obedient to), we need a universe that's smooth at large scales but coarse at small ones ("small" being a relative term here). In other words, we need to explain the simultaneous and partially contradictory existence of both a sky dotted with stars and a background of uniform microwave radiation.

Indeed, one of the main tools in our cosmological arsenal, inflation, just makes the situation worse—at least at first blush. One of the most important jobs of inflation was to solve the so-called horizon problem by ensuring that everything in the universe had enough time to coordinate to get everyone looking the same before *whooshing* out and making the big giant cosmos that we're all used to by now.

So if inflation is a super-homogenizer, a primordial smoothie blender, how can there still be chunks of fruit floating around and getting stuck in the straw?

The answer is . . . inflation. Yup, at it again, with a major trick up its quantum mechanical sleeve. The same hammer that cosmologists use to solve our scientific shortcomings when it comes to understanding the large-scale properties of the universe also explains the structure contained within it.

When I first introduced inflation a million years ago, I talked about it in half-circumspect terms, like I wasn't really believing what I was telling you. And if I remember correctly, I promised I would return to inflation and provide some better evidence for it. And here we are.

When inflation was first dreamed up by Alan Guth in the 1970s, he didn't really have the puzzle of structure formation in mind. Astronomers hadn't yet made deep enough extragalactic surveys to map out significant structures in our universe, and the formation of stars and galaxies was a problem for other people. In fact, for a while cosmologists were seriously considering the possibility that cosmic strings, weird and twisty fractures in space-time itself related to magnetic monopoles, might provide the mixing necessary to stir up some density differences in the early universe.[1]

Alas, it was not to be.[2] Instead it was the humble *inflaton*, the highly hypothetical quantum field that drove inflation before the close of the first second. You see, it didn't just inflate space-time, fulfilling its cosmological duty to solve all sorts of nasty artifacts of the early big bang picture. It had a bonus side effect: it inflated *everything*.

Remember those quantum fields that permeate all of space-time? And how we can never look at individual particles the same way again? I know it was just a chapter ago, but it was pretty heavy stuff. The world around you is infested with these writhing, contorting, vibrating quantum fields, and even in their ground state—the state, by definition, of minimum energy—they support a massive amount of energy.

If you take a box and empty out all the particles and radiation, you have an empty box. But the quantum fields soak into space-time itself, so your box is still full of them. They're just not "active," with no major waves on them, so there are no particles bouncing around between the walls. But ground states don't mean zero-energy states. That's a rookie classical physics mistake, and you need to start thinking like a quantum mechanic if you're going to make it in this universe.

Even in their ground state, the quantum fields are abuzz with activity, humming just below the level of perception. They normally don't affect the physics of our everyday world or even the subatomic realm: all our normal interactions happen "on top of" this ground state, so we don't notice any difference. It doesn't matter if you build your house on the seashore or on top of a mountain—if you want to get to the second floor, you still have to climb a flight of stairs.

But we know this "vacuum energy" (one of the names we give to this ground state of the quantum fields) exists because it does occasionally interact with our world. If an atom absorbs a bit of radiation, causing an electron to jump to a higher energy level, what makes that electron want to come back down? The higher energy level is perfectly stable as long as the electron isn't jostled. Well, it's the vacuum energy, the background hum of the quantum universe, which provides the necessary jostling to knock the electron off its ledge and back down to a lower-energy state. There's also the Casimir effect, where two parallel metal plates will attract each other in a vacuum, due to the quantum fields on the outside overwhelming the ones on the inside and creating an inward pressure. It's a bit too complicated to fully explain here but serves as a needed reminder that vacuum energy is indeed a thing.[3]

I'm going to switch metaphors now. Instead of a background hum, I want you to think of vacuum energy as a background *fuzz*. You might be tempted to call an empty patch of pure space perfectly smooth and featureless, but the quantum fields, sitting in their ground state, provide a subtly textured backdrop to the painting of our universe. If you zoom in on the sub-sub-sub-microscopic scale, all the way down to the Planck length, you'll see that space-time isn't flat—it's bumpy, constantly roiling and boiling with these quantum motions. The features in space-time are driven by the effervescent hum of the fields: they are constantly changing their energy state, and the addition or subtraction of energy, no matter how fleeting, affects the space-time around them.

When inflation inflated the universe into ridiculous proportions in the first second of the universe, it *also* inflated these quantum fluctuations, enlarging them out of the realm of the microscopic and into the world of the merely small.

Throughout the radiation era, through the collapse of atoms in recombination, and into the dark ages of matter domination, these seeds, these wrinkles in density, persisted and grew. What's more, new *perturbations* (introducing the technical term because that's how we roll) began entering the observable universe. This is an important part of the story of inflation and how we know it's (relatively) accurate, so here comes another metaphor.

Let's say you're in a big concert hall, listening as the orchestra plays, I dunno, the Brandenburg Concertos, or that opening theme from *Game of Thrones*. The room is full of sound, from short-wavelength high-pitched flutes to long-wavelength low-pitched brass. In an unexpected flash, the concert hall undergoes a phase transition and experiences inflation, growing exponentially larger in a fraction of a second. The walls of the room, the orchestra itself, and even your friends get carried away from you. The hall is now much larger than the limit of what you can see—you're completely isolated in your observable patch.

Along with the concert hall, the sounds emanating from the instruments were stretched too, most of them reaching such incredibly long wavelengths that they don't fit inside your observable bubble. In essence, they're frozen, too big to resonate in your universe and be heard. But over time your bubble grows, continuing to expand. As it grows larger, it catches up with the frozen-in sound waves, and as soon as the waves fit, they vibrate the air molecules and you can hear the sound again.

One by one, tone by tone, frequency by frequency, you're able to piece back the last moment of the orchestra, starting with higher frequencies that fit easily in your bubble, then over time the lower and lower pitches. Each successive note perturbs the air in a different wave, resonating at longer and longer scales.

The act of inflation pulled some fundamental quantum fluctuations outside our observable universe, but as our cosmos expanded it "caught up" to continually larger fluctuations. Once they (jargon mode on) "entered the horizon" (jargon mode off) they could begin influencing the motion of matter and energy.

And we can see the influence of those quantum fluctuations, inflated well beyond their usual scale, like a disturbing macro photograph. The enlarged fluctuations left an imprint on the cosmic microwave background, where minute density differences, planted in that first instant of the big bang, persisted through the first few hundred thousand years of cosmic history. Those density differences were small but detectable. By the 1990s, detailed measurements of the cosmic microwave background revealed temperature differences of one part in ten thousand.[4]

More missions followed and flew, and today we have incredibly high precision maps of that relic light.[5] We're finding a universe that's homogeneous—the cosmic microwave background is essentially the same no matter the patch of sky—but not completely so. There are tiny, subtle differences in temperature, and the statistical properties of those variations match precisely what we expect from this inflationary picture. While we can't see behind the great firewall of the CMB, the cataclysmic events of previous epochs rippled and shook through the young universe, leaving their stamp on that first flash of light.

The pieces of evidence that indicated an inflationary paradigm for an early universe took decades to emerge after the initial accidental detection of the CMB by Penzias and Wilson in the 1960s. That's because these temperature differences are incredibly small, and you need to measure good chunks of the sky before you get reliable enough statistics to claim a result. Hence science at an industrial scale is needed—no more of this amateur-hour business. But after enough seriously hard work with space-based missions such as COBE (the Cosmic Background Explorer), WMAP (the Wilkinson Microwave Anisotropy Probe) and Planck (just . . . Planck), nature eventually revealed the secrets of the primordial universe as written on the face of the microwave background.

Notice that this will be a common theme running forward—of discoveries and observational insights coming less and less from individual astronomers or quick-thinking theorists, and more and more from teams of dedicated professionals with armies of grad students laboring underneath them. There's a social commentary on the state of modern science somewhere in there, but I won't pursue that (in *this* book).

The end result is this: It seems, by all lines of available evidence, that something like inflation occurred in our past, and that the event of rapid expansion enlarged fundamental quantum fluctuations to macroscopic scales. And these small differences in density and temperature planted the kernels for much larger structures.

All it takes is a little gravity and a lot of time.

In our universe, with interactions governed by gravity at large scales and long times, the rich get richer and the poor get poorer. If there's a random spot with just a tiny bit more density than average, it will have a slightly stronger gravitational pull, and nearby matter will be attracted to it. As more matter piles on, the density difference grows larger, and the corresponding gravitational attraction becomes stronger, encouraging more nearby neighbors to join the party.

So over time, over the course of hundreds of thousands of years, stretching into the millions, pockets of gas in the universe began to condense and grow larger, and the large spaces in between them slowly, steadily emptied out. The process may have been seeded by exotic quantum forces, but by now it was driven by the entirely benign and predictable gravitational pull.

There was movement in the dark.

Slowly, silently, stealthily, throughout the era of these dark ages, long after the light from the cosmic background had shifted below the visible, a web of matter began to weave itself. Starting from the microscopic, clumps of hydrogen and helium atoms arose, accumulating and growing steadily larger, forming dense clumps a light-year across, tens of light-years across, thousands and millions of light-years across. The first major structures in our universe began to form.

All in darkness, in a cooling and expanding cosmos, over millions of years the slow but relentless process of gravity drove matter to collect in on itself, generating small-scale differences set against a large-scale backdrop of a homogeneous universe.

And in one small corner of the universe, in a pocket of gas no different from any other, the densities reached a critical point. As matter accumulated on itself to form larger structures, the internal compressions deep in their cores grew and grew, eventually reaching a point where the gravitational pressures overwhelmed the electrostatic repulsion of the hydrogen atoms, forcing them together in a nuclear embrace, igniting a fusion reaction.

The first star was born. The cosmic dawn had arrived.

The first stars to appear in the universe are thought to be massive beasts, easily a hundred-plus times more massive than the sun. We think they arrived on the scene around the time that our home was celebrating its hundred-millionth birthday, less than a hundredth of its current age. Compared to the exotic, cacophonous pace of earlier stages in cosmic evolution, events take an achingly slow time to wind up in the older, colder, slower cosmos of more recent eras.

In the sweet joys of astronomical conventions, these stars are known as Population III (or simply "Pop-3" if you want to sound fluent in astro-jargon), even though they're the first generation. Stars like our own sun are called Population I, with the intervening generation Population II no matter the ordering scheme. These first stars are something special—they are made of pure, unrefined hydrogen and helium, without a trace of heavier elements.

The first stars were the first intense sources of light to appear in the cosmos since the dense, hot plasma that filled the universe before recombination. One by one, stars began to ignite across the cosmos, flooding its volume with an intensity of light that hadn't been seen for millions of years.

While I'd love to give you more details, the formation of the first stars is shrouded in mystery. Part of the problem is observational—as far we can tell (and we've looked really, really hard) there are no Population III stars remaining in our present-day universe. Since bigger stars lead shorter lives due to the increased ferocity of their nuclear reactions, that's a clue that they must have been heavy beasts. It's only with our deepest surveys that we even catch a glimpse of the first galaxies, which formed eons after those first stars.

Those young galaxies may—emphasis on the *may*—contain a few retirement communities of the first stellar generation residing in them, so some of the combined light that makes its way across the vastness of time and space into our telescopes could show a trace of that older population. In recent years there have been tantalizing hints, but nothing supremely conclusive.

The other roadblock to knowledge of the first stars is theoretical. Earlier epochs in the universe were satisfyingly *linear*, which in scientific circles means it's easy to calculate and make predictions from the math. It's well-behaved and housebroken. You can solve problems the good old-fashioned way, with pencil and paper. Even inflation itself can be understood, at least at the broad-brush level, on the chalkboard. The predictions for the temperature of the cosmic microwave background were calculated by hand. The growth of density perturbations under the influence of gravity, the addition of new disturbances as the universe grew larger, our model of the young universe for the first millions of years, the lot of it, are all shaped by relatively simple physics and straightforward math.

But stars are not linear creatures. Complex flows of gas and matter are not easily solved with a piece of chalk. Predictions for the relationships between matter, radiation, and nuclear forces are not for the faint of heart. Understanding this epoch requires serious horsepower in the form of sophisticated computer simulations that attempt to recreate the physics of these early epochs. It's a tough business, recapitulating the universe in silicon, and usually requires a wealth of observational data to constrain and restrict the theoretical models. But with a lack of observations, the simulations are leading the way, with theorists tweaking input parameters and models of physical processes to glean some sort of useful information about those first stars.[6]

So despite the breaking of the cosmic dawn, we're still relatively in the dark when it comes to first formation of stars. The bigness of the stars is assumed to be generally correct, though, because of the nature of the clumping of material in these protogalaxies (there wasn't a lot of mixing going on that could steal away energy) and because direct hydrogen fusion is relatively inefficient, so you can pile *a lot* of stuff on before it kicks in. But estimates vary wildly—simulations suggest anywhere from forty to three hundred times the mass of our sun.[7]

It's tough to say at this early stage (in understanding) what happened at this early stage (in the evolution of the universe). But we do know that stars formed relatively early on in our cosmic history, and that it led to another, perhaps final, transition.

For the first time in a long, long time, there was a new source of energy. The newborn stars flooded their surroundings with high-energy photons the likes of which hadn't been seen since the previous, plasma-filled epoch. Until those stars formed out of the murky depths of cold and slippery hydrogen, the only meaningful source of radiation was the quickly dimming and redshifting afterglow of the cosmic background. While those numberless photons could still be felt, they didn't pack nearly the punch they used to—their influence over the lives of matter had long since waned.

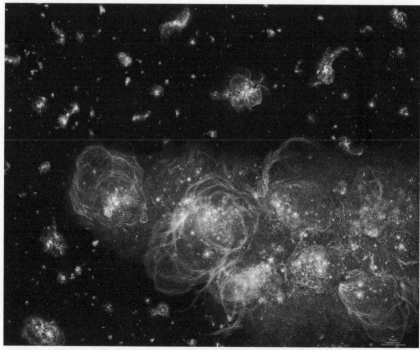

In its early, heady days, the universe was a violent place. Here, still-forming protogalaxies crash into each other, igniting a round of brilliant star formation and presumably some spewing of gas. (Image courtesy of NASA.)

But these new stars? With potentially gargantuan sizes reaching hundreds of times the mass of the sun? They were powerhouses of light and brilliance. What's more, those first stars began clumping together. The process of structure formation laid down in the inflationary era didn't stop with the ignition of nuclear furnaces in the hearts of those first stars—they were simply the heralds of greater things to come.

With the first few hundred million years, either soon after the formation of the first stars or concurrently with them (we're not exactly sure), the first galaxies began to collect themselves. The great "island universes" first confirmed by Edwin Hubble were already on the cosmic scene at a relatively early time. The progenitors of what would eventually evolve into our own Milky Way or our neighbor Andromeda had already begun coalescing, studded with sparks of massive stars.

The first galaxies were probably small and relatively dim. They hosted intensely luminous stars, but not many of them. They hadn't had billions of years to accumulate new materials from the surroundings. They didn't have efficient pathways for mixing gas and dust (and "dust" wasn't even really a thing back then) to produce batches of hundreds or thousands of stars in a single go. But they did, as far as we can tell, already host black holes.[8]

As with everything else in this epoch, we're not exactly sure how the first black holes formed. Nowadays black holes form from the deaths of massive stars, when the explosive nuclear energy in their cores can no longer support their own crushing gravitational weight. And—hey!—there are tons of giant nuclear gasballs floating around in the cosmic dawn. So it's easy to suspect that they'd eventually die, spectacularly, and leave behind a population of black holes with masses a few dozen times the mass of the sun, or thereabouts.

But we're not seeing evidence for these ancient small black holes, the same kind that fat stars leave behind every single day. We're seeing evidence for much bigger monsters: the supermassive black holes. The width of a solar system, these hulking beasts lurk in the hearts of galaxies and tip the scales somewhere north of a few tens of millions times the mass of the sun. They're thought to grow from mergers of many smaller black holes, plus the occasional gorging on surrounding interstellar gas. They're very common in the present-day cosmos; we think *every* galaxy hosts such a black heart, even the Milky Way.[9]

The name for ours is Sagittarius A*, for the morbidly curious.

Of course it's hard to see black holes directly, either in our neighborhood or in the distant reaches of the early universe. They're black, space is black—

not a lot of contrast. But when one of these black beasts is actively feeding, when gas is collecting and swirling around its great maw, the extreme gravitational compression of the material itself causes it to shine fiercely. These are the most powerful engines in the known universe, easily outshining a supernova. We have a few names for them—active galactic nuclei, quasars, etc.—but they all mean the same thing: when black holes feed, the material they consume screams in blazing agony.

And we see them in the young universe. Really young, when the cosmos isn't even a billion years old. That's strange; the biggest black holes that can power such emissions take multiple generations of stars to be born, live, die, leave behind a remnant, then have multiple remnants collect in the galactic center and merge. That takes time, right?

In fact, all the characters in this act of the play—the stars, the black holes, and the galaxies—all seem to come in much more quickly than we expected. The universe is showing signs of a certain maturity despite its youthful age.

Is the formation of complex structures just faster than we normally think? Did it occur faster in the early universe for some reason we haven't fathomed? Are there other forces at work? Perhaps the initial massive black holes condensed out of primordial hydrogen straight away, bypassing the conventional route of stellar death. Perhaps these clumps seeded the early, rapid formation of those young galaxies as well. Maybe the universe just acted different when it was a kid, despite operating under pretty much the same physical principles as today.

Who knows. As of the time I'm writing this, it's an active area of research. The physical dark ages may have ended with the ignition of the first stars, but when it comes to human knowledge of the first billion years, we're still lacking a lot of light.

Like I said, the biggest challenge to studying this epoch—the dark ages and the cosmic dawn—is just that: there isn't a lot of light. Up until the recent uses of neutrino detectors and gravitational wave observatories, the only way for us to access the universe was through electromagnetic radiation. Light, in all its glorious forms. Different kinds of light, like the infrared that Herschel accidentally found, or X-rays, ultraviolet, or radio, reveal the full variety of physical processes afoot in our cosmos.

When it comes to the dark ages, nothing's alight. And even though the cosmic dawn is defined as the moment when stars arrived on the scene, (a) there aren't a lot of them, and (b) they're stupid far away. What's most irritating for the modern astronomer is that we have so much information *around* this era. At the early-universe side, we have the cosmic microwave background, a flood of primordial photons, up there amid the most well-studied celestial object in human history.

At the modern-universe side, we have, well, the universe. Stars and galaxies aplenty, buzzing with activity, glowing in every radiation band conceivable. A noisy, messy, chaotic universe. It perplexed our astronomical ancestors, but lately we've become rather good at building gigantic surveys to map out large swaths of the local cosmos in one go.

But the dark ages? There's simply not enough light to see that far (at least with current instruments—upcoming missions like the James Webb Space Telescope should be able to pick out some of the first creatures in our universe).[10] It would be completely, utterly hopeless—a case where the observers just walk up to the theorists and shrug—if not for one thing, a small quirk of quantum mechanics.

Neutral hydrogen, the kind that inhabited the universe after the recombination event (making neutral hydrogen was, I guess, the entire point of recombination), is as simple as simple atoms can be: a single proton and a single electron. Those two particles have mass, charge, and spin. Remember spin? If you don't, go reread chapter 7 right now.

Welcome back. Left to its own devices, hydrogen likes to have the proton spin one direction and the electron spin the other—that's its preferred lowest-energy ground state. Of course every once in a while, the hydrogen atom can get a kick, flipping the electron over so that it spins in the same direction as the proton. Since this state just has a teensy-tiny bit more energy than the true ground state, it's actually quite stable. In fact, there usually should be no reason at all for the electron to flip back around. It just has no quantum mechanical incentive to do so.

But another gem of quantum mechanics is that if you wait long enough in the subatomic world, unexpected things can happen. Energy barriers can be broken. Walls can be tunneled. Particles can appear where they please. And if you give a neutral hydrogen atom a few million years to think about it, it can spontaneously flip its own electron over and settle back down into its usual up-down spin state.

That means a release of energy from one quantum level to another. That means there's emission of radiation with the exact same energy. That radiation is deep in the radio—around 1400 megahertz. Put your hands out in front of you and pretend you're holding a basketball. I'm serious: put down this book and pretend you're holding a basketball. I don't care if you're in public. We need to do this together.

Thank you.

That's about twenty-one centimeters, the wavelength of radiation emitted by neutral hydrogen via this subtle quantum mechanical trapeze act of the electron. We routinely use this so-called 21-cm radiation to map out pockets of neutral hydrogen in modern-day galaxies, and it might just give us a window into the dark ages. The severe challenge is that this primordial signal is weak, just the barest whisper in the deep background, and it's shouted over by all the other nearby sources of radio emission. That said, as I write the hunt is on for this peek into our distant past using massive radio telescope arrays. Even if a confirmed, reliable detection is made soon, though, it will be many years before a solid analysis can be performed.[11]

The study of the dark ages and cosmic dawn is perhaps the last frontier of modern cosmology, the great unknown and unmapped realm between the nearby galaxies and the cosmic microwave background itself. That said, we can still manage to make progress. We're trying our best, OK?

The young universe was filled with neutral hydrogen, from metaphorical corner to metaphorical corner. But look around the universe today; as I hinted at just above, there are only *pockets* of neutral gas remaining. Indeed, most of the hydrogen and helium in present-day galaxies is in the form of a plasma. The sun's a ball of plasma. The interstellar gas is a plasma. The material that floats around and between galaxies is a plasma, a superthin yet still hot plasma.

We *know* that the gas in the young universe just after recombination was neutral. We *know* that the gas in the present-day universe, 13.8 billion years later, is not. How did it get *reionized*, and how the heck are we supposed to figure it out?

Perhaps it was that first generation of stars, massive and angry, hot enough to spew out ultraviolet radiation. That kind of light packs enough punch to knock an electron off an atom. Perhaps it took a lot more than that; maybe it was the supernova death of those stars that injected enough energy and X-ray radiation into their surroundings. Possibly it was the combined might of millions of stars—the first galaxies—to transform the cosmos. Probably those massive black holes, forming surprisingly early, did the trick. As gas fell into their horizons, it compressed and heated, releasing torrents of radiation in the process.

Whatever the cause, about a billion years into its history, the universe underwent its last major metamorphosis, an era we call the Epoch of Reionization. Slowly, starting from pockets of stars and protogalaxies, the final veil was lifted in the universe as neutral hydrogen and helium were reenergized and reverted back into their plasma state, a state they have retained until today.

As with everything else in the dark ages, we don't know a lot about this process. We knew it *had* to happen, because we know the state of the universe before and after this transformation. And while we don't know the details yet, we do know that this transmutation was messy, ugly, and awkward.

After a billion years, the universe hit puberty.

Compare yourself at ten years old and twenty years old. Chances are you're a *radically* different person. Compare yourself at twenty versus thirty. You're probably a little rounder around the middle but otherwise unchanged. That's what puberty does—it transforms you from a kid into a protoadult. The Epoch of Reionization was when the universe finally grew up. Done were the days of wild phase transitions and late-night plasma parties. In were the days of mortgage payments and bad backs.

In the dark ages, the universe was threaded with a dense, warm soup of neutral gas. If you could transport yourself there, it would look totally unfamiliar and alien, as strange as any of the earlier epochs. After the cosmic dawn and reionization, the universe looked like . . . the universe. Vast, dark, transparent, and dotted with galaxies swirling with billions of glittering stars. Much like you since your twentieth birthday, it hasn't physically changed much since then, at least in comparison to younger days.

After a billion years, after the last of the neutral gas had been swept away and left to cower in small pockets inside galaxies, after the cosmos had once again been filled with light and heat and warmth, the greatest structure in the universe began to coalesce.

CHAPTER 9

OF MATTERS DARK AND COLD

Fritz Zwicky knew that something fishy was going on.

It was the early 1930s, and in the years following Hubble's spectacular and surprising result—that galaxies are *things* and that they are, on average, redshifting away from us, implying that we live in an dynamic, expanding universe—astronomers had undergone a quick change in heart. With gigantic new telescopes like the hundred-incher on Mount Wilson, which Hubble used as a massive eye to peer beyond the limits of the Milky Way, astronomers went from debating the very existence of extragalactic objects to racing to catalog as many of them as possible.

If you can't beat 'em, join 'em, I guess.

Now in survey mode, astronomers reclassified known spiral nebulae as spiral galaxies (although the term "nebula" would still persist for some time, thankfully we've been able to drop that anachronism by now; otherwise our discussion of cosmology would be even more saddled by the presence of yet another historical-jargony albatross) and started mapping the heavens for as many deepest-sky objects they could find.

Zwicky, the Swiss astronomer renowned for his acumen, creative (and sometimes crazy) thought processes, and prickliness, took a special interest in a group of galaxies in the direction of the constellation Coma Berenices.

Now, even the most ardent astronomy enthusiasts will admit that there's basically nothing interesting happening in the constellation. Heck, even its name, *Berenice's Hair*, doesn't exactly inspire the wonder and majesty we typically associate with the night sky. There are a few Messier objects, but otherwise it's an unremarkable patch of darkness.

Unless you look deeper. The universe in that particular piece of the sky is swarming with galaxies, hundreds of them, much more than other random directions in the sky. That itself is an intriguing fact to note, which astronomers in the first half of the twentieth century surely did: Galaxies outside the Milky

Way are not scattered around randomly and uniformly. No, there are vast empty patches and what appear to be *clusters* of these galaxies. The astronomers at the time had no idea of what to make of this, but as we've already begun to see in the previous chapter, the solution will become apparent to them soon enough.

Zwicky diligently mapped, measured, and cataloged as many galaxies in one cluster in particular, the Coma Cluster, as he could; a veritable cosmic butterfly case of extragalactic creatures. All those galaxies had roughly the same distance from our home, which was the first clue that it wasn't an accident of optics that led to their clustering in the sky. No, these galaxies were associated with each other in deep space—they lived together.

Since the galaxies lived together, Zwicky guessed that this cluster must be stable. If you saw a random group of people at a random time of day at a random house, you might guess that those people aren't total strangers collected together by pure chance—they're probably a family. They've been together for a while. Sure, it *might* just be a guest-filled house party on a Thursday afternoon, but it's not likely. This argument is from statistics, and it's a pretty useful one in cosmology. Structures that don't last long (in cosmological terms) simply won't persist long enough for us to catch their light in the small window of time that we've been observing the deep heavens.

It's a safe assumption: if you see something, it's already been there a long time.

Zwicky cleverly used that assumption to start answering a very simple but deeply difficult question: how much does that cluster of galaxies weigh? How massive is it? It's one of the most straightforward questions we can ask of anything, even celestial objects. It's high on the standard list of questions for describing anything, really. You get a baby announcement, and you get a few key pieces of information: name, sex, length . . . and weight.

Astronomical objects aren't gendered, so as soon as we name something (like, say, the Coma Cluster), our next job to break out the rulers and scales. Which is hard, because things in space are typically gigantic and far away. But Zwicky used an old physics trick known as the *virial theorem*. Originally developed in the nineteenth century, it connects kinetic to potential energies within a system of particles bound together.

We'll jump right to Zwicky's application of the theorem to explore what all that jargon means. First off, the "particles" are going to be entire galaxies. Yes, galaxies are gigantic, but they're peanuts compared to the size of a cluster. The kinetic energy is related to the speed of every galaxy—the faster the gal-

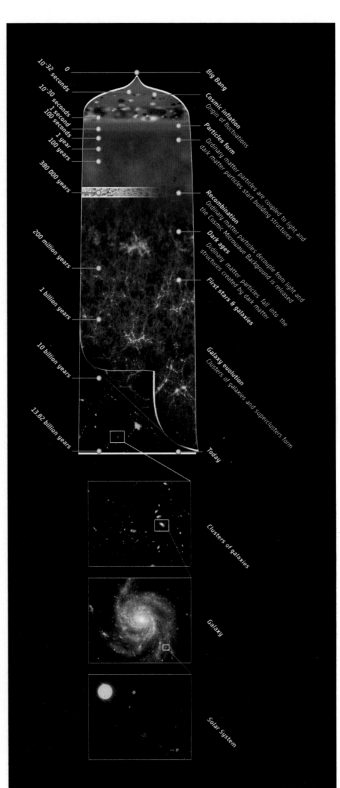

10⁻³² seconds — 0 — **Big Bang**

10⁻³⁰ seconds — **Cosmic inflation**
Origin of fluctuations

1 second
100 seconds
1 year
100 years — **Particles form**
Ordinary matter particles are coupled to light and dark matter particles start building structures

380 000 years — **Recombination**
Ordinary matter particles decouple from light and the Cosmic Microwave Background is released

200 million years — **Dark ages**
Ordinary matter particles fall into the structures created by dark matter

1 billion years — **First stars & galaxies**

10 billion years — **Galaxy evolution**
Clusters of galaxies and superclusters form

13.82 billion years — **Today**

Clusters of galaxies

Galaxy

Solar System

A peek of what's to come in the universe, even though we're just getting started. The scale is chosen to give each major transformational epoch its rightful due. *Image courtesy of NASA / JPL / Planck Collaboration.*

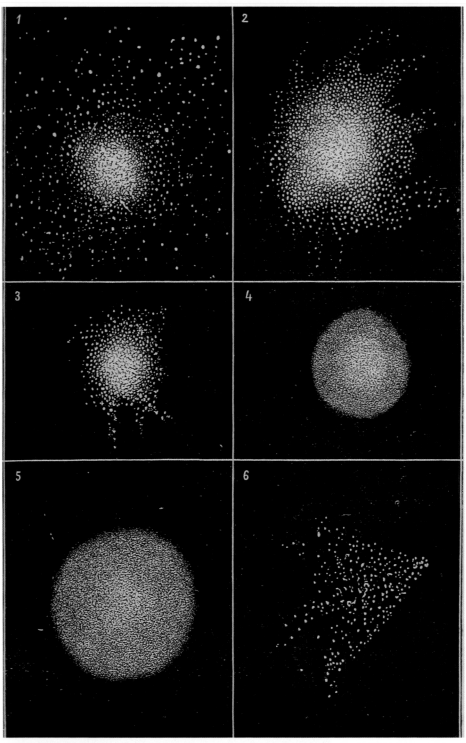

In the 1800s, Sir Norman Lockyer sketches a bunch of fuzzy things in the sky, some quaintly called "nebulae."

We used to think that the Andromeda Galaxy was just a "nebula." Whoops. *Image courtesy of NASA / JPL-Caltech.*

It turns out that there are hundreds of billions of these "nebulae." Oh, boy. *Images courtesy of NASA / ESA.*

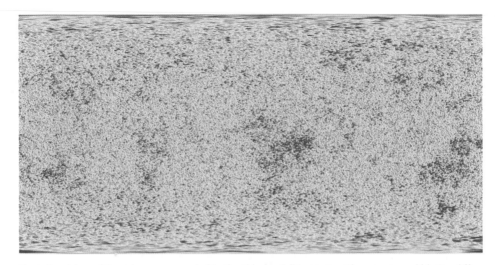

The baby picture of the universe as revealed by the Planck satellite mission. These are incredibly tiny differences in microwave temperature—less than one part in 10,000 variations away from the average, which are directly related to density differences way back then. The small blips are the seeds that will one day grow up to become galaxies; the large blotches are caused by sound waves crashing around the infant cosmos. The image is distorted at the top and bottom because it's representing the entire sky (a sphere) as a rectangle, just like how on a map Greenland and Antarctica look way bigger than they actually are. *Image courtesy of ESA / Planck Collaboration.*

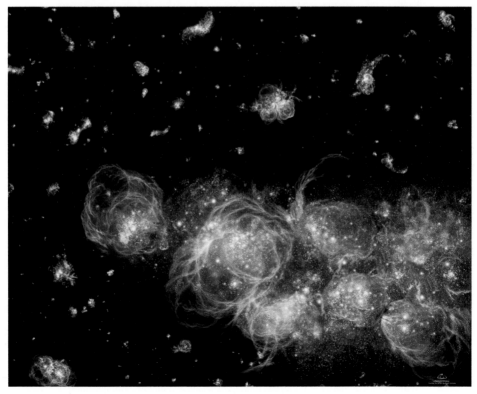

When two galaxies collide: a hot mess. *Image courtesy of NASA.*

The (in)famous Bullet Cluster viewed in the many ways necessary to reveal the lurking dark matter. In this image, the relatively tiny galaxies are embedded in the much larger clusters, which are caught in the act of a titanic collision. The hot gas of the clusters is tangled near the center, while most of the mass—including dark matter—has sailed through itself. *Image courtesy of NASA / CXC / CfA / STScI / ESO WFI / Magellan / University of Arizona.*

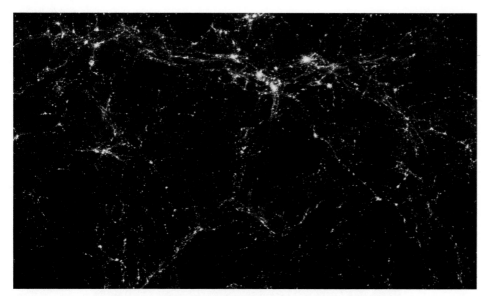

While immense for us humans, this is but a small slice of the vast cosmic web. It is taken from a computer simulation of the universe soon after it began coalescing into the large-scale structure that we know and love today. We can see in this 50-million-light-year section the dense and tangled system of clusters and filaments embedded among the open maws of the voids. *Image courtesy of Wikimedia Creative Commons; author: Andrew Pontzen and Fabio Governato; licensed under CC BY 2.0.*

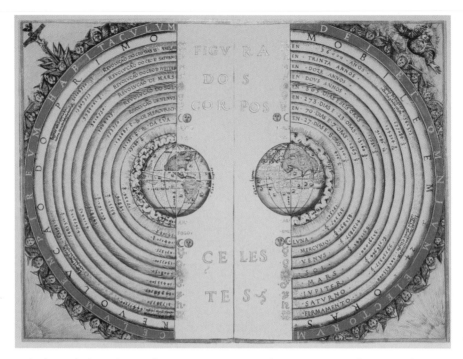

We've learned a lot in the past few centuries, I swear it. The top image is an illustration of the old-school geocentric model of the universe, drawn by Bartolomeu Velho in 1568, just on the eve of the coming cosmological revolution. The bottom is a creative interpretation of our modern view, where, due to the finite speed of light, different observers see themselves at the "center" of "their" universe, with greater distances revealing a younger cosmos. As in the timeline shown earlier, the scale is chosen to highlight various scales. . . . I'm guessing old Bartolomeu had a similar plan. *Image courtesy of Wikimedia Creative Commons; author: Pablo Carlos Budassi; licensed under CC BY-SA 3.0.*

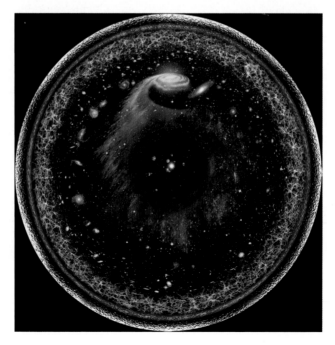

A type Ia supernova expends more energy than its entire host galaxy. Briefly, but it counts. *Image courtesy of NASA / ESA.*

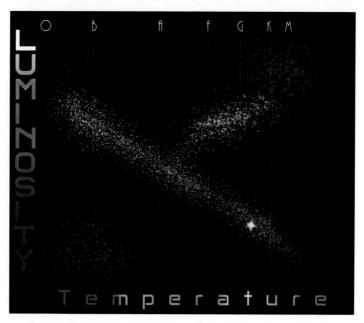

An example of a so-called H-R Diagram, which neatly sorts stars according to their temperature and their luminosity. How wonderful: a pattern emerges in nature, unlocking a clue to stellar lives. *Image courtesy of Wikimedia Creative Commons; author: Jessica Repp; licensed under CC BY-SA 4.0.*

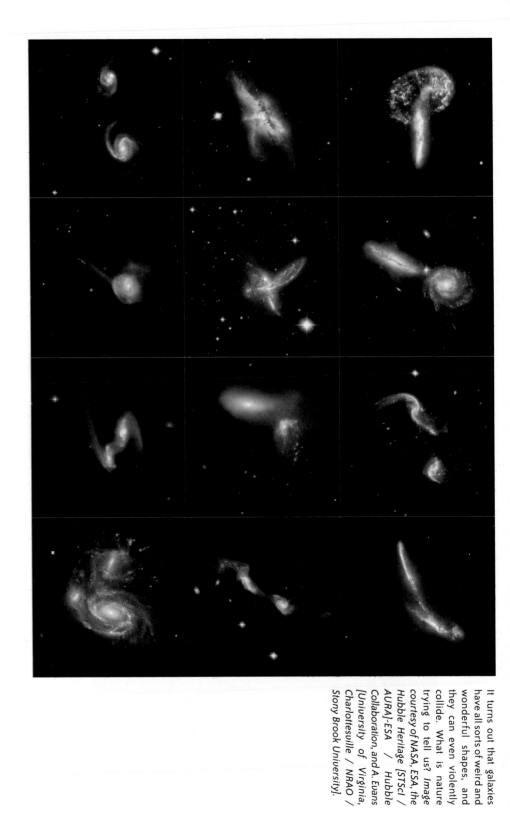

It turns out that galaxies have all sorts of weird and wonderful shapes, and they can even violently collide. What is nature trying to tell us? Image courtesy of NASA, ESA, the Hubble Heritage [STScI / AURA]-ESA / Hubble Collaboration, and A. Evans [University of Virginia, Charlottesville / NRAO / Stony Brook University].

axies are buzzing around, the greater the kinetic energy. The potential energy here is provided by gravity—the mutual interaction of all the galaxies provides the glue holding them together.

This is where stability matters. The galaxies are pulled together by gravity, by the invisible strings (or deformations in space-time, if you're feeling relativistic) that insist on bringing objects closer together. Resisting that is the intrinsic velocities of the galaxies themselves. You can imagine a cluster of galaxies like a tremendous swarm of bees. Gravity pulls the swarm into the space of a sphere and really, really wants to make that sphere smaller, but if the bees have enough energy, they'll just keep buzzing around inside that sphere.

If gravity overwhelmed the kinetic energy, the cluster would have collapsed a long time ago. If the galaxies were too energetic, it would have exploded. Since the time needed to collapse or explode is short (again, just as a reminder, that's "short" compared to things like the age of the universe), it's a good guess that these forces are in balance.

This allows a person like Zwicky to make a relatively easy measurement (the average speed of the galaxies is given by their redshift) and convert that into a relatively hard measurement (the mass of the cluster). So he did.[1]

He got a number that was a bit too large, honestly. He knew how much mass was in all the stars surrounding the solar system, and he knew how much light they produced. So that gives a pretty simple formula: if you see *this much* light, then you can make a good guess about how much *mass* is generating that light.

But that number, based on counting all the hot, glowy stuff on the Coma Cluster, didn't agree with the number produced by the virial theorem. The virial theorem was based on simple kinematics; it was just motion that was affected and balanced by gravity, which was on pretty solid footing. But that calculation, based on physics as good as anybody's, resulted in a cluster mass five hundred times larger than you would have guessed by adding up all the light sources.

Uh, what? What could explain this discrepancy? Maybe stars in the Coma Cluster are weirdos and don't behave like any other star anywhere else. That doesn't seem to fit. Maybe the laws of physics are different way out there than over here. Kepler surely wouldn't appreciate that, but he's dead and can't complain, so we'll keep that in our back pocket. Perhaps the Coma isn't in equilibrium after all—maybe we caught it in a really awkward time in its life and the ground rules of the virial theorem don't apply. Maybe, but while possible, that's unlikely.

Maybe it's something else. Maybe there's material in the Coma Cluster that isn't all hot and glowy. There could be macroscopic or microscopic *stuff* that we simply aren't seeing. In the words of Fritz Zwicky himself, maybe the cluster contains vast amounts of *dunkle Materie*.

It's German, and I'll translate for you: dark matter.

It was about time for a shake-up anyway. While astounding and capable of bending even the biggest of brains, the observational results of Hubble coupled with the theoretical insights of Einstein and company at least made sense when taken together: we live in an unfathomably large universe that's getting fatter every single day. We may not *like* that answer, but at least it's an answer.

But Zwicky's dark matter wasn't an answer—it was a question. Other clusters were tagged and bagged by later astronomers, and they each showed this curious disagreement between different measurements of the mass. But while it was noted, nobody made a serious move to try to explain it. I can't blame anyone: if you were an astronomer interested in cosmology, there was all sorts of excitement about the big bang, the cosmic microwave background, and the creation of elements to pass your time. No need to worry about tiny niggling issues like giant discrepancies in cluster mass estimates, right?

And so Zwicky, and everybody else, moved on from the mystery of *dunkle Materie* to other, more pressing and interesting problems. A generation went by without much further thought on the subject. Then in the 1970s another astronomer, Vera Rubin, saw the same problem crop up in another situation, and what was considered a forgotten yeah-we-should-get-around-to-that-someday problem rose from the dead, setting the stage for one of the great enduring mysteries of the modern cosmological age.

The brief window of explaining and understanding the universe in a consistent, coherent story was quickly closing. Oh well, it was nice while it lasted.

Rubin was studying the motions of stars in other galaxies, which by any measure (and especially mine, since I'm writing the book) is an amazing feat of human curiosity. Just a century ago, we were struggling to identify the distances and motions of the stars right in our own galactic neighborhood, and by the late 1960s, Rubin was investigating, with great success, detailed interior movements in structures millions of light-years away.[2]

It's a pretty straightforward measurement once you get the hang of it. Pick a galaxy. Zoom in on various parts of it. Look at the spectrum. Identify some known elements. Measure the shifting of the spectral lines associated with those elements relative to what you know on the Earth. The same song and dance from a century before repeated over again, just at unimaginably large distances.

In the case of the entire galaxy, you can apply this technique to find that the whole structure is moving, and usually you'll find that it's in a direction away from us. That's what Hubble found. But if you focus your scope on individual bits of the galaxy, you'll pick up some extra motions on top of the general movement. If it's rotating (and, uh, they do) then one side of the galaxy will be slightly blueshifted—spinning toward us—while the opposite side will be a little bit redshifted—spinning away from us. You can repeat this exercise at different spots in the galaxy and build up what's called a rotation curve: the orbital speed at various distances in a galaxy away from its galactic center.

I won't hold back the plot twist: Rubin didn't find the speeds she was expecting. I know, shocking.

Here's the deal. When you just look at a spiral galaxy, and I really mean *look*, you'll notice that there's a big bulge of material (stars, gas, etc.) in the center, with a relatively thin and empty disk surrounding it. And our dear old friend Kepler can tell you what the orbital speeds of stars out in the disk ought to be. OK, fine, it's really Newton's universal gravity, but it was Kepler who first spotted the harmonious motions in the planets, and universal gravity being universal and all, the same method applies on these gargantuan scales. The speed of an orbiting object, whether a planet in the solar system or a star in a distant galaxy, depends on its distance from the central gravitating body and the mass of that body. A bigger sun = faster planetary speeds. Closer orbits also = faster planetary speeds.

Great, lovely. If we apply this to a galaxy, where most of the material is *obviously* concentrated in the center, then we ought to find that stars nearest the center orbit the fastest, with a general laziness setting in as we move farther out in the galactic disk.

But alas, no. Rubin saw something completely different: totally flat rotation curves. Stars out in the boondocks, at the very edge of civilization, were orbiting just as fast as their more centrally located cousins. The stars in galaxy are simply moving too fast. There isn't enough gravity to contain motions of that speed—the grand, beautiful spirals that we know and love should have flung themselves to pieces long ago, not stayed glued together for eons.

And this wasn't just the case in one galaxy, but across dozens. So, unlike with Zwicky's cluster, we can't appeal to a chance fluke of observations. Even if one of these galaxies were simply acting strangely, most of them are in equilibrium. Apparently, moving too fast for the gravity of your own galaxy is just the natural state of things.

And in the decades since Zwicky's initial proposal, we had much more confidently pinned down the relationship between stellar output and mass. Even accounting for the mass of nebulae, whether they glowed brightly or not, there wasn't enough *stuff* inside a galaxy. Once again, one measurement (counting all the hot and glowy bits) was disagreeing with another measurement (one based on motions). Zwicky saw it in a single cluster, and the world ignored it. Rubin saw in galaxies everywhere she looked, and the world started paying attention.

We're at a crossroads. Nature is not playing fair. Different measurements of the same quantity—answering the simple question of the total mass of a galaxy or cluster—were revealing different answers. And not just by a little bit. At best, the amount of glowing matter in a galaxy or cluster, even accounting for all the wave bands from radio to gamma rays and all the possible sources from nebulae and stars to brown dwarfs and giant black holes, was about one-fifth the value required by other mass estimates.

Rubin's result was as simple and annoyingly counterintuitive as Zwicky's: either our laws of physics don't work at galactic scales, or there's a dim and/or invisible component to their recipe. What is nature trying to tell us? New physics or dark matter? With only the results of Vera and Fritz (now *that's* a sitcom) to work from, we can't tell the difference between the alternatives.

We've been in this situation before, with a set of observations contradicting what we expect. It's kind of how science advances, so at least this is familiar territory. And in the case of gravity, we've taken both paths in the past. For example, Mercury. Tiny little Mercury, the closest to the sun and the most swift-footed of the planets. It has the most elliptical orbit of all the planets (not counting the dwarfs like Pluto and Eris), and the point where Mercury comes closest to our sun lazily traces out a circle over the years. Most of this motion is due to the gentle but persistent gravitational tugs of the outer planets, but a small part of that motion couldn't be explained by the gravity of Newton.

Perhaps it was another planet in our solar system, an inner-inner world of fire, nicknamed Vulcan, orbiting close enough to the sun to remain hidden to the ancients, only making its presence known by gravitational flirtations with Mercury. But searches turned up empty. The answer here was new physics: one of the first clues Einstein had that he was on the right track was that general relativity could fully explain Mercury's orbital oddities.[3] In other words, the portrait of gravity as painted by Newton breaks down close to the sun, and it takes a revolutionary new view of the cosmos to understand what's going on.

So maybe the quandary presented by Rubin and Zwicky could be explained by new physics. Maybe general relativity, just like Newtonian physics before it, can't cut the mustard past a certain scale. We love you, Albert, and your theory is a thing of beauty. But maybe it's just not good enough, pal.

But to be perfectly honest, the Vulcan approach has worked in the past. The orbit of Uranus was also behaving oddly, given the known denizens of the solar system and our Newtonian knowledge of gravity. Instead of modifying Newton to account for the observations, astronomers instead posited the existence of a new planet to explain the curious orbits in the outer solar system, and in due time the *agent provocateur* was found—the planet Neptune.[4]

In that case, it wasn't new physics but previously unknown matter that best explained observations.

But what to do with galaxies and clusters? Well, when nature fights dirty, fight back—with science.

In the decades since Rubin's reinvigoration of the dark matter debate, astronomers around the world have embarked on an all-out observational war, measuring, comparing, and testing galaxy after galaxy and cluster after cluster with as many methods as humanly (and, in the age of more advanced technology, robotically) possible.

I'm going to be polite here and warn you that I'm about to absolutely inundate you with more evidence for the existence of dark matter. It's not that I'm trying to beat this concept into your brain, but—no, wait, that's exactly what I'm trying to do. The reason is that the mystery of dark matter persists to the present day (at least, to the time of me writing this book). We have tons of rock-solid evidence that something funny is going on out there in the great expanse, but we're much more hazy on what's causing it.

Because of this, there's a lot of fear, uncertainty, and doubt among the general public (surely not *you*, but maybe someone you know) when it comes to understanding this facet of our universe. In the past few centuries, we've solved a lot of mysteries of the heavens above us, but continued observations have revealed deeper, perhaps more sinister machinations in the motions of celestial objects. One of them is the nature of the earliest moments of the big bang, when forces were so extreme and exotic that we have trouble even theoretically navigating them.

The other is here, in the old and cold universe of today. It's not some relic of the distant past, a problem that we can leave on the doorstep of future generations of scientists, safe in the knowledge that our overall picture is coherent. Dark matter is present *today*, in galaxies and clusters all around you. And when I get there, you'll see that it's probably inside you too.

We've already seen Zwicky's realization that *dunkle Materie* might be a thing inside clusters of galaxies based on the motions of galaxies whizzing around inside them. But threaded between those galaxies is an incredibly hot (up to a hundred million Kelvin—that's hotter than the core of the sun) but incredibly thin (about one thousand particles per cubic meter; compare that to the 10^{25} air molecules per cubic meter that you're breathing right now) plasma. It's so thin that the particles—protons and electrons—travel for about a light-year before interacting with each other, but when they do, they emit X-ray radiation in a process known as *Bremsstrahlung.* It's German for "braking radiation," and it may or may not be one of my favorite words in physics.[5]

So the gas is hot and emits X-rays. The hotter it is, the more X-rays come out; ergo, we can measure the intensity of X-rays and estimate the gas temperature. And just as with the galaxies themselves, there's a connection between the temperature of the gas and the total mass of the cluster. If the cluster is in equilibrium, then the gas can't be too hot (or the cluster explodes) or too cold (or it collapses). Since that is obviously not happening, we can get a rough handle on cluster masses. Result: not all the cluster mass is visible.

Here's another one. Let's rewind all the way back to the first dozen minutes of the universe, when all the protons and neutrons were hitching up to form the universe's initial supply of hydrogen, helium, and lithium. The prediction of the abundance of those light elements, spawned from our understanding of nuclear physics, was (and is) a triumph of the power of the big bang model. Those same calculations place a hard upper limit on the total amount of baryonic (if you remember your jargon lessons, that's a particle like a proton or neutron) material available in the universe. When we compare that to all our cluster observations, we end up with a number that's about one-fifth too shy. So in principle, we're measuring the same thing, the mass of the universe, but at different times (in the first minutes versus the latest billion years). Those numbers ought to be the same, because where is all the mass going to go? But instead the clusters are much fatter than they should be, given the elemental building blocks available to them.

Check this one out. Remember that story of inflation and the early gravitational growth of structures? It's a good story and worth remembering. There's one big caveat that I deliberately neglected to mention, because I wanted to save it for this moment: in order for the galaxies to have the sizes they do, we need a form of dark matter. Specifically, a kind of dark matter that doesn't interact with light. The problem with pure baryons is that they get easily distracted. Gravity tries to pull them together, but the intense radiation pulls them apart, a process we saw play out with agonizing repetition until matter and radiation finally went their separate ways with the birth of the cosmic microwave background.

But by the time of recombination, a few hundred thousand years into the history of the universe, the seeds were already planted . . . by dark matter. Some form of invisible matter could freely ignore radiation, pooling itself together in those early years, creating the nests that baryons would eventually collect in to start building larger structures. In other words, even though we can't see it, the cosmic microwave background and the later structures to emerge depends on the presence of dark matter.

Given only baryonic processes operating in the early universe, there simply wasn't enough time to build galaxies, including the Milky Way. Including you. Yes, you. Without dark matter, high-density structures (and I'm not calling you *dense*, per se, just noting your density relative to—never mind) couldn't have formed.

Not convinced yet? Let's take a look at gravitational lensing. Matter, whatever its form, will bend space-time, which will deflect that path of light. We've tested this like crazy. So if you're looking at, say, a gigantic galaxy cluster and wondering about its mass, you can use the bent light from *background* objects to figure it out. If you look at a distorted image through a regular glass lens, and you know what shape the undistorted image ought to have, you can use your Knowledge of Physics to figure out the properties of the lens doing the distortion.

We know what galaxies look like because we have, well, a lot of samples. So when light from a distant galaxy passes through a not-so-distant cluster, that image gets distorted, and we can compare it to what galaxies look like without a funhouse mirror and Sherlock out the properties of the cluster doing the lensing. You know, its mass.

So that's a completely totally different method for measuring the mass of big cosmological objects. And guess what answer it gives?

Yup: large objects in the universe are more massive than they appear at first glance.

I don't know about you, but I'm getting the feeling that nature is trying to tell us something.

The evidence has been piling up for decades now, and our options for explaining that evidence—what nature is practically shoving in our faces every time we go to ask the very simple question of how much a galaxy or cluster weighs—have severely narrowed in the time since Rubin's, and especially Zwicky's, initial results.

Can all these combined results—galaxy rotation curves, cluster masses, the cosmic microwave background, early-universe physics, the mere existence of galaxies, and gravitational lensing—be explained by new physics? Is Einstein not enough? It's always a possibility, but as the years go by, it becomes increasingly slim.

The first serious attempt came shortly after Rubin's observations with a theory called MOND (once again, I'm not in charge of naming things). Short for Modified Newtonian Dynamics, it's exactly what it says it is—it modified Newton's fundamental laws so that the relationship of mass, acceleration, and force isn't what we're used to.[6] The key is that everything appears perfectly normal on the surface of the Earth or in the solar system, but once you get to galactic scales, it breaks down and has to be, well, modified.

This works great for explaining galaxy rotation curves, because it was explicitly designed to work great for explaining galaxy rotation curves. No invisible matter here, just different physics at work. But in order to make a theory capable of coherently explaining both galactic and early-universe observations, it needed to be elevated to be fully relativistic—in other words, it had to be written in a way similar to special relativity (which, dark matter or not, nobody was arguing against). That's the game we have to play in cosmology: if you're not relativistic, you don't have enough explanatory oomph to take you everywhere you need to go.

The result of marrying MOND with relativity is called TeVeS, for Tensor-Vector-Scalar. I won't go into the messy details, but the short version is that it's largely ruled out: TeVeS predicts certain results for gravitational lensing and the cosmic microwave background that don't agree with observations.[7]

The real bullet that finally killed modified laws of physics was detailed observations of the appropriately named Bullet Cluster.[8] That cluster of galaxies is itself a train wreck (though not to be confused with the Train Wreck Cluster, which is something else), where two massive clusters slammed into each other long ago, and we have a pretty picture of the wreckage. When we look at this system with different techniques, like a team of forensic investigators trying to understand a murder, we get a complete picture of what's going on.

First, the galaxies. Galaxies are like buzzing little bees in the giant volume of their host clusters. When clusters collide, it's like two swarms of bees headed in the same direction: for the most part, they just sail on through without interacting. So the galaxies end up on opposite sides of where they started. Fine, nothing surprising there.

Next, the hot, thin gas between the galaxies that fills up the bulk of the cluster. As thin as it is, it still can't help but get tangled up with its counterpart in the opposite cluster during the merger event. And when those two giant balls of gas slam into each other, we get all the rich and glorious physical interactions that we've come to expect with interacting balls of hot gas: cold fronts, shock waves, instabilities, the works. When we examine the Bullet Cluster with X-rays, we see all the fireworks happening in the center of the interaction, with the two sets of galaxies sitting on opposite sides, safely navigating the merger event unscathed.

Now, where's the mass? We have exquisitely good gravitational lensing maps of the Bullet Cluster, showing not only how much total material the colliding pair hosts but also where it's distributed in space. How handy is that? Those lensing maps reveal a curious pattern: the concentration of mass is not tangled up with the hot gas in the center but is more closely associated with the galaxies. Even then, though, it's not mapped directly to the galaxies themselves but, rather, smoothly distributed throughout the remnants of the clusters.

That hidden mass is much, much larger than can be provided by the galaxies alone.

There's no picture of modified gravity that can sufficiently explain what we see with the Bullet combined with every other observation we have of the universe. I know, I know. It would have been super awesome to have a handle to lever ourselves up and past Einstein's relativity. A couple of Nobel Prizes would have been tossed around, and we'd be working on the Next Big Challenge.

Oh well. Like I said, nature isn't playing fair.

So Einstein gets to stay in the game, but as I said earlier, something has to give. The universe must have some additional, previously unknown component.

Maybe it's normal stuff, just dim. Like dead or failed stars, rogue planets, or black holes. Those don't give off a lot of light (obviously), and with a bit of finagling and just the right circulation, they would have all the right properties for galaxy rotation curves, lensing, and all the rest. But the only way to make black holes is through the death of massive stars. And the only way to make botched stars is to have giant clouds of star-forming material that failed to ignite fusion. To have the *dark* matter really be *dim but otherwise normal* matter, we need a lot of . . . normal matter, and that's ruled out by our knowledge of the early universe, like the processes that build the first nuclei and the growth of the first galaxies.[9]

There are, of course, a bunch of half-baked, and sometimes quarter-baked, ideas floating around in academic circles, and it would be exhausting to give an exhaustive review. Give me five theorists, and I'll walk away with six theories on dark matter. I wanted to insert this caveat because of a profound and—dare I say—noble sense of completeness, but I'm not going to spend a lot of time on them because they're usually pretty dang awkward solutions to the dark matter problem, and not as fleshed out and agreeably comprehensive as the one I'm about to present.

Which is WIMPs. That's right: WIMPs. Astronomers, as we've seen, have a flair for the ridiculous acronym. Weakly Interacting Massive Particles. The main historical competitors to WIMPs were the MACHOs—the MAssive Compact Halo Object, the name given to the chunks of "normal" matter, like failed stars, that might have explained the observations but never quite did.

So our best solution to the dark matter problem is called a WIMP, and we're just going to have to live with that.

To explain what the heck a WIMP is and why you should care, I need to mention one other property that dark matter, whatever it is, has to have: it has to be *cold*. That means that at the time it comes on the scene in a big way, it has to have speeds much lower than that of light. The reason is structure. If the invisible part of our universe is too "hot," then it has a much easier time ignoring the effects of gravity, and this "washes out" smaller structures like galaxies. Since galaxies very much exist, the dark matter has to help rather than hurt the formation of structure billions of years ago, which means it has to be amenable to the flirtatious whisperings of gravity.[10]

Thus particles like the neutrino, which absolutely inundate the universe and (this turned out to be a surprise) have a little bit of mass, aren't a good candidate to be the dark matter—they're too hot, and if you added too many of them into the recipe of the universe, enough to point to it and say, "That's the dark matter—we've known it all along!," the resulting pastry will be too bland and smooth, lacking the delicious, flaky layers of our modern cosmos.

I've been dancing around this issue up until now, so I'll get right to it: as best as we can tell, the explanation for the dark matter problem in our universe is a new kind of subatomic particle previously unknown to science. Something that doesn't interact with light, probably doesn't even interact with itself, and is cold.

And our best theoretical candidates for that particle is the WIMP, a (P)article that is (M)assive and (I)nteracts only via the (W)eak nuclear force. Why this one? Well, cosmologists aren't the only people on the planet hunting for never-before-seen particles. Those crafty high-energy theorists are constantly churning out idea after idea, coming up with ways to break the chains of the standard model of particle physics and extend our knowledge of the subatomic realm. The initial forays into the uncharted territory of new physics predict the existence of a host of new particles, just as Dirac's musings led us on the path to antimatter.

These predicted particles just so happen, by sheer happy coincidence, to have the right properties to explain the dark matter as we see it in the universe, and especially to have the right properties in the early universe to form the seeds of the kinds of structures that we observe at large scales today.[11]

Of course there are many untested routes beyond the standard model, and these routes differ in their predictions of what *precisely* the dark matter might be—if there is even a single particle responsible for all that mass, and not a family. But the good news of WIMPs is that these hypotheses make testable predictions, so we can at least *try* to rule them out.

If dark matter is truly a WIMP, then each and every galaxy is flooded with them. The same way you are constantly surrounded and bombarded by radiation, which you notice, and neutrinos, which you don't, trillions upon countless trillions of WIMP particles are zipping through your room right now, through the hands holding this book and through the brain thinking these thoughts. You're *drowning* in dark matter.

You don't notice because—as per the definition—it only interacts via the weak force, which is very rare. We don't notice the effects of its gravity on small scales like the solar system because it's relatively smooth. Gravity only cares about *differences* in density, so you have to get out to supergalactic scales to notice any substantial variations and the interesting gravitational properties that come into play.

Right now, as I type and probably as you read, there are detectors all around the world dedicated to hunting for a dark matter particle, hoping to catch enough rare chance interactions to confirm a detection of the dark matter particle. So far, they've turned up empty, which has helped rule out some models and tighten the noose on the exact properties of the WIMP.

While they're still the most appealing candidate for the dark matter, the attitude toward WIMPs has taken a more skeptical bent as of late, especially after the Large Hadron Collider smashed some particles together to look for paths leading away from the standard model. Unfortunately, those searches have—so far—turned up empty, and the simplest and easiest paths, some of them containing a WIMP candidate, appear to be dead ends.

That said, there are a few other candidate ideas floating around the journals and blackboards of academia that predict new particles without them being strictly WIMPy, so there are still a lot of uncovered stones.

Again bowing to that noble sense of completeness, I need to mention that dark matter as an explanation for our cosmic observation does have some weaknesses, dealing with the detailed structures of galaxies and the numbers of satellites a galaxy might have. Those are, of course, topics of active research and vigorous debate.[12]

Disclaimers out of the way, this is how we do science. You pick the explanation that has the fewest assumptions and is able to explain the most observations. It's a very simple strategy that has uncovered and unlocked mystery after mystery in the cosmos. And despite our best efforts to develop a viable alternative, it appears that we live in a universe dominated by an exotic, cold, invisible form of matter.

This particle, if it exists, lives outside the standard model. Nothing we know of—not the neutrino, not any of the quarks or mesons or whatever-ons—can explain the cosmic observations. *Something* funny is going on in the universe at large scales, and that funny something is deeply related to our unquenchable thirst to understand physics at its deepest level.

Whatever that particle is, whatever its true nature and intent, it is by a wide margin the dominant player of the matter game in our universe, beating our familiar light-loving baryons at least five to one. So when you enjoy a clear dark night dotted with countless sparkling stars, you're looking at the mere representatives of cosmic structures. All the particles you know and love, the ones that constitute the physics of the familiar, of the warmth of sunlight on a summer's day, of the solid rocks underneath your feet, of the exchange of ions in your blood vessels, are a minority.

Stars and even galaxies are lighthouses on a distant, hidden shore. A beacon of light, signaling the presence of larger masses, tracing their outlines without revealing more. The journey started by Zwicky and Rubin, and continuing to the present day, leads us to an inescapable and uncomfortable conclusion: we do indeed live in a dark, cold universe.

CHAPTER 10

THE COSMIC WEB

T he cosmic web is insultingly big. To state the bare fact that it's the single largest pattern found in nature does supreme disservice to the word *largest*. If there were a superlative to express a quantity greater than the greatest, that adjective might come close to addressing the magnitude of the cosmic web.

Think of the largest thing you possibly can. A planet? The Earth is so large that even though it's round, it appears flat in your backyard. A solar system? NASA's New Horizons probe traveled at thirty-six thousand miles per hour and took nine and half years just to make the hop to Pluto. A galaxy? A bustling stellar megalopolis, home to hundreds of billions of stars and a hundred billion suns' worth of gas.

The cosmic web is *made* of galaxies, the same way that your body is made of cells. But even that metaphor breaks down—as do all metaphors—when describing the cosmic web. The cosmic web is made of galaxies, the same way your body is made of cells . . . if your cells were a million times smaller than they are.

But the galaxies themselves are only representatives of the true bones of the web: dark matter. The matter in our universe appears (or doesn't—ha, sorry, cosmology joke) to be made of some nonluminous particle or family of particles, one that doesn't interact with light or really anything. It's the dark matter that began pooling together billions of years ago; the "normal" matter simply fell into the already-existing gravitational valleys. When we see a galaxy, we must imagine a "halo" of dark matter surrounding it; likewise for a gigantic cluster. When we see the web, the light-emitting galaxies are tracers of the true structure underneath. Thus while everything I'll talk about below concerns stars and galaxies, keep in mind that those are only the metaphorical tips of the cosmic icebergs.

Let's start with some baby steps and work our way up. Voyager 1, launched in the late 1970s on a grand tour of the outer planets of the solar system, finally penetrated the bubble of our sun's influence, as defined by the boundary where the stream of charged particles racing outward from the sun's surface begins to mix with the general galactic milieu, in 2012. In three hundred years, that little spacecraft—which is no bigger than a small car, mind you—will reach the inner boundary of the Oort cloud, a thin, diffuse shell of frozen debris left over from the formation of the solar system.

Voyager 1 now enjoys the privilege of being the only humanmade object in interstellar space, the long gulfs of emptiness between the stars that make up the Milky Way. It will eventually pass by another star, coming within 1.6 light-years of Gliese 445, an unremarkable red dwarf currently situated about 18 light-years from the sun.

In forty thousand years.

While it's difficult to predict exactly, astronomers are pretty sure that's the closest Voyager 1 will come to another star. Ever.

In 230 million years, traveling at a steady thirty-eight thousand miles per hour, it will complete a single circumnavigation of the Milky Way without meeting anything larger than a stray bit of dust.

Here's another perspective. The sun's nearest neighbor is Proxima Centauri, another unremarkable red dwarf (they're rather common) about four light-years from our home. If you were to build a scale model of our galactic neighborhood, and you were to put the Earth a scant three feet away from the sun, Proxima Centauri would be two hundred miles away.

And we're just getting warmed up.

The Milky Way galaxy itself is around one hundred thousand light-years across. Simple math reveals that you could fit twenty-five thousand sun-Proxima distances across its breadth. In our scale model, with the sun three feet from the Earth and Proxima two hundred miles from that, the Milky Way in its entirety would stretch five million miles, which would put the edge about twenty times farther than the moon.

That's a big model.

Let's revisit our exploration of the deep universe using the handy new phrasing we learned when we first encountered Hubble and his fantastic result: the *parsec,* the quintessential astronomical jargon word. Like this: the Milky Way is about thirty thousand parsecs, or thirty kiloparsecs, across. It's still unimaginably gigantic, but at least I don't have to type a bunch of "illions" anymore.

The Andromeda Galaxy, the nearest major neighbor to our own galaxy, is about a megaparsec, or million parsecs, away from us. That simple statement hides an intriguing fact. Although galaxies are tens of thousands of times bigger than their constituent solar systems, the distances between galaxies are only a few times larger than galaxies themselves, making them sort of close together. Relatively speaking.

But the most interesting thing about the large-scale structure of the universe on scales bigger than even galaxies, and the reason it has a name like *cosmic web*, is that galaxies aren't just arranged randomly throughout the cosmos, a fact that quite surprised the early cosmographers—the mappers of the universe.

But in order to reveal the structure of the universe, you have to go much larger than our local patch. The work of Hubble and others to establish the expansion of the universe relied on a comparatively nearby sample of galaxies—not nearly enough to reveal anything other than a scattered smattering of so-called spiral nebulae. Sure, there were dense beehives of galaxies like the nearby Coma Cluster, which flirtatiously suggested the existence of greater structures, but for decades cosmologists weren't sure if that was a significant feature or just a random collection.

But beginning in the 1970s, astronomers began to perfect the technology necessary to systematically survey the locations of galaxies far from the familiar, using a potent combination of improved telescopes and computerized search algorithms. It was like the classic duo of scope + camera that transformed our understanding (or lack thereof) of the nineteenth-century universe, but on steroids. The new surveys pushed both wider and deeper, creating the first-ever maps of our universe on a truly universal scale. Galaxy by galaxy, insignificant point of light by insignificant point of light, each dot representing a hundred billion warm nuclear hearths set against the deep, vast coldness of our cosmos, structures began to appear.[1]

Long, thin ropes. Broad walls. Dense knots. Immense voids.

It was a web. A web made of galaxies. One by one, survey target by survey target, astronomers realized that at the largest scales, the galaxies of our universe are not scattered like salt sprinkled on a table. There was a pattern. There was order. The cosmic web is, and I'll repeat myself in case you missed it earlier, the single largest pattern found in nature. It stretches from one end of the observable universe to the other—and most likely beyond that, too, though we'll never see it.

The complexity of the web is astounding. The same substructures and intricacies you would find in a spiderweb are found at these vast scales. The filaments are long and tenuous, in some case only a few galaxies wide but stretching for dozens of megaparsecs. Some walls are almost impossibly large. The first one was simply called the Great Wall, only a few megaparsecs thick but hundreds of megaparsecs across.[2] It was soon given a more specific name (the CfA2 Great Wall, for the curious, named after the survey where it was discovered) because it turns out there are a bunch of Great Walls in our universe.

The walls and filaments intersect at the clusters of galaxies, the largest gravitationally bound structures in the universe. Home to a thousand galaxies or more, they are relatively compact—a bare megaparsec or two from edge to edge. Most clusters are isolated creatures, settling into hydrostatic equilibrium and spherical symmetry long ago, but a few are cosmic car accidents, giving clues to their formation and constituents (there's a gruesome analogy here that I won't pursue further), like the infamous Bullet Cluster. Inside the clusters we find not just galaxies but also a hot, thin gas: the tenuous plasma of hydrogen, weak enough to register as a vacuum in the laboratory but hot enough to emit X-rays.

And between all these great extragalactic structures sit the voids. By volume, most of the universe is void—the filaments, walls, and clusters are tight agglomerations of galaxies, with high density but low volume. By contrast, the voids are almost completely empty, literally devoid of much matter at all. They're not totally empty, though, and detailed surveys of individual voids reveal a surprising feature.

If you zoom in on a portion of the cosmic web and look at the voids, you'll see a few dim dwarf galaxies. It's only by careful, detailed, prolonged observations that these galaxies reveal themselves to even the most dedicated astronomers, but they are there. You can play the same game that you would do for the entire cosmic web: map out their positions. And when you do that, you get a surprising answer—a faint echo of a cosmic web, embedded inside the voids.

The structure of the universe isn't quite a fractal, even though fractal cosmology has a long and storied history, but today for various reasons that word is the ultimate f-bomb in cosmological circles. But parts of the cosmic web do resemble a fractal, at least superficially; the cosmic web can be seen as a series of *nested* cosmic webs, each "level" dimmer and smaller than the last.[3]

This has only come to light—astronomy pun again, sorry—in the last few years. That's because astronomical surveys are set by brightness limits. You design a telescope, say, *this* big (I'm holding out my hands so you can see what I'm saying), with a sensor capable of detecting *this* many photons in each pixel. That sets a hard limit on the dimmest thing you can take a picture of.

In deep astronomy, objects can be dim because they're far away, or they can be dim because . . . they're dim. So the same technology that allows us to sweep farther into the distant reaches of the cosmos allows us to scan in detail the emptiest patches, looking for any signs of light.

It's only once you reach small enough scales—around five million parsecs—that the interactions between individual galaxies ruin the harmoniously nested picture. Also, once you reach *large* enough scales—around one hundred million parsecs—the cosmic web loses its distinctiveness. At those scales, the breakdown of cluster, filament, and void becomes meaningless. A single hundred-megaparsec patch of the universe looks pretty much the same as another hundred-megaparsec patch. Each patch will contain its own unique arrangements of structures, an intergalactic fingerprint pattern, but the statistics of those structures (like the number and sizes of voids, the typical distance between clusters, etc.) become the same.[4]

That is the scale of true cosmological uniformity. Remember the cosmological principle, the ground-base assumptions used in the mathematics of general relativity to describe and understand the history of the universe? Those mathematics obviously don't apply in the solar system or even the galaxy—there's too much other stuff going on, like cosmic rays and chemistry, that dominates the interactions at those scales.

But out past a hundred megaparsecs, all those complicated, messy, stubborn physics just blend together into a seamless blob, revealing the bland uniformity to satisfy the cosmological principle that our universe is homogeneous and isotropic. The cosmic web is huge and magnificent, a pattern stretching from one end of the universe to the other (ahem, the universe doesn't have an end, but the imaginative language was too juicy to pass up). The observable universe is about ninety billion light-years, or thirty gigaparsecs, across, and it's filled with this frightening, beautiful web of galaxies.

There's one more pattern in the cosmic web that I want to tell you about. It's a subtle one, but the detection of its very existence was an important—and recent—milestone in cosmology, and if we've learned anything in the hundreds of years of scientific astronomy, it's that subtlety is the name of the game.

Let's briefly zoom back to the early universe, before the decoupling of matter and radiation and subsequent unleashing of the cosmic microwave background. The universe was a hot, dense plasmatic soup of particles, crashing around violently in the tense bath of radiation that permeated the cosmos. In that primordial sauna, the first instabilities grew and would eventually lead to the grand structures that we observe today.

There were also sound waves. Yup, *sound waves*. Any medium, even a plasma, can support the existence of rippling waves of pressure, which is exactly what sound waves are. The early universe was a sonic cacophony of profound, booming, bone-quaking vibrations. The constant competition between matter and radiation for dominance in the young plasma created those sound waves, and they continually crashed throughout the universe.

That is, until matter and radiation went their separate ways. Once that divorce happened, the sound waves were left hanging, frozen in mid-plasma. A sound wave is a pressure wave, and the peak of the wave represents a region of higher-than-average density. Just like sound waves in air: when someone speaks, her larynx vibrates air molecules, pushing them together, creating a rippling effect of over-and-under densities that travel outward; eventually, those alternating densities of air molecules vibrate your eardrums.

Imagine if you could freeze a moment in time and see the sound wave hanging in the space between you and the speaker, with some air molecules smooshed closer together and some spread thinner. That was the state the atoms were left in just at the moment of recombination, 380,000 years into the history of the universe. It was a small effect, naturally, but it persisted, leading to a quiet but persistent bias in the grand arrangement of galaxies that would follow over the course of the upcoming billions of years.

This feature, known as baryon acoustic oscillations (baryon = normal matter; acoustic = like a sound wave; oscillations = wiggles), is detectable, quite clearly in fact, in the cosmic microwave background as large patchy blobs of temperature differences. And it's detectable, quite subtly, in the organization of galaxies at the very largest scales. Those frozen-in sound waves sit today at a scale of around 150 megaparsecs—an ever-so-slight "bump" in the average density of galaxies; a peculiar feature in the cosmic web; the largest structure in the universe carrying a birthmark from its primordial origins.[5]

I lied. The most fascinating thing about the large-scale structure of the universe, the arrangement of galaxies on the grandest of scales, isn't the fact that it resembles a weblike pattern; it's that it *moves*.

The cosmic web is alive. Not living-creature alive, but alive in the sense that it was different in the past and will be different in the future. The cosmic web has *evolved*, and its history is still being written. The pattern isn't static. Galaxies are buzzing around in orbits with the clusters. The filaments are really freight trains, ferrying loads of galaxies from the intergalactic rural countryside to the clusters. The walls are collapsing. The voids are growing.

Despite the frightening scales of the cosmic web, the physics that govern its evolution are ridiculously simple. It's just gravity. And time. Loads and loads of time.

The seeds were first laid down when the universe was less than a second old, and we've watched them grow like a nursery of baby chicks over the course of the first billion years, from random microscopic fluctuations to the first dense clumps, stars, galaxies, igniting in an explosion of energy at the cosmic dawn. But the process didn't end with the formation of the first lighthouses. Gravity does what gravity does, and slowly, over time, ever-larger structures arrived on the scene.

After the seeds were set, initially in a dark universe but then in a light-loving one, all you needed to do was wait a few billion years and voilà! Your very own cosmic web. Gravity and time—the recipe couldn't get simpler.

The iconic weblike pattern owes its existence to gravity, with its seeds laid down by the event of inflation itself. The same microscopic fluctuations that gave rise to the speckling of the cosmic microwave background eventually grew up—it's no wonder that I like to refer to that afterglow light pattern as the baby picture of the universe.

We live in what's called a *hierarchical* universe. The biggest objects—filaments, walls, and clusters—grew from gluing together smaller structures. As soon as galaxies assembled, they collapsed into distinct groups called, well, groups. The groups piled themselves into the walls and filaments and then streamed along those highways into the clusters to join their friends and their busy lives.[6]

But this process didn't happen all at once. Some galaxies have already completed their journey to their local cluster and have settled in nicely. Others are only recently getting a move on. Hence we get the weblike patchwork of in-process structure formation that we see in the universe today.

So it's fair to say that the cosmic web *condensed* out of the early, uniform primordial soup. Neat.

The best part is that we get to literally see this process, which played out at a glacial pace over the course of billions of years, frozen into the very pattern of galaxies that we observe with our deep surveys. That's because the science of astronomy is more closely related to paleontology than physics (please don't tell the astronomers I said that).

Light takes time to travel. It's fast, but not *that* fast, and especially not that fast compared to the cosmological distances it must travel to reach our eyes and optics. Our own sun is eight light-minutes away from the Earth. The photons that are making you squint or tanning your skin on a summer's day at the beach left the surface of the sun eight minutes ago. If the sun were to wink out of existence—a catastrophe, to be sure—we would at least have eight minutes of blissful ignorance before darkness fell upon us.

The measure of a light-year doesn't just mark out distances; it really does quantify a duration of time. A nameless star twinkling in the sky isn't just unfathomably distant from us in space, it's not *present*—the stars are neither here nor now. A couple thousand light-years to that speck of light means that we're seeing the star not as it is at this snap of my fingers but as it *was* thousands of years ago. Our image of the great Andromeda Galaxy isn't at the same age as the Milky Way.

As we probe ever deeper with our telescopes, we slowly reveal the *younger* universe. Surveys of the "nearby" universe (cosmologically speaking) exhibit the beautiful, intricate lacework patterns of the cosmic web. More distant galaxies trace out structures in an intermediate state of formation, and the faintest objects we can capture with our telescopes only hint at the beginnings of larger organizations. Past that are the dark ages, unlit by stars, which we are only beginning to probe. Once those are unveiled, we'll surely see a more primordial state (uh, at least I hope so).

At the very limit of our observations, we must switch to microwaves to map out the cosmic background radiation, the fossil relic from an infant cosmos.

The ancients viewed the universe from a geocentric point of view, with Earth at the center, surrounded by concentric spheres holding the moon, sun, and planets, surrounded by the uppermost celestial sphere hosting the not-so-distant stars themselves. In a strange quirk of perspectives, the time-capsule effect of light-travel time delays produces a picture of the universe with the Milky Way at the center, surrounded by a contemporary cosmic web, a hazier proto-web around that, a ring of blackness when the universe was plunged in darkness, and the thin shell of the primordial cosmic microwave background surrounding everything.

The universe has no center and no edge, but because light takes time to hop from its source to our telescopes, our *view* of the cosmos ends up recapitulating the models that we strove so hard to reject. I *swear* this is a coincidence—or at worst, a deep-seated psychological need to place our fragile humanity in some important position or focal point, lest the overwhelming largeness and complexity of the universe drive us mad.

Structures in our universe were built from the bottom up, from smaller objects to larger, grander cathedrals of light. Let's see how this process plays out in our local patch of the universe. To start, you should know that there's a place in our cosmos called the Great Attractor, and we're headed right for it.

Remember those maps of the ancient cosmic microwave background? And how tiny differences in the temperature from one spot to another reveal the face of the early universe? And how those differences pile up on each other to form the grandeur of the cosmic web? Good.

Before cosmologists could make *that* map, they had to subtract out our motion relative to the background itself. The Earth is orbiting the sun, the sun is orbiting the center of the Milky Way, and the Milky Way itself is buzzing through empty space. Altogether, it adds up to a brisk six hundred kilometers per second. Not too shabby.

This motion was first detected in the CMB itself, which is slightly (and "slightly" here means 0.0035 Kelvin) warmer on one side than the other due to the redshift of our motion. But we knew the speed of the Earth's orbit (thanks, Kepler!) and the speed of the sun's orbit in the great Milky Way. Add those together and it's . . . well, not six hundred kilometers per second. Ergo, the galaxy moves.

But if our home galaxy is moving, where is it moving to? Well, for one thing, we're on a collision course with the Andromeda Galaxy. That million-parsec space between our neighbor and us won't stay that comfortably large for long. In about five billion years—give or take a billion, depending on your exact definition of "collide"—we will collide with Andromeda, forming a single beefy galaxy yet to be named. Andromeda Way? Milkedromeda? The options are endless.

But that inevitable galactic headlong rush *still* isn't enough to explain the mystery of our intergalactic motion. Where exactly are we going?

As (bad) luck would have it, this is not an easy question to answer. For decades, astronomers have been carefully mapping the cosmos, starting with our local environs and steadily moving outward. But we can only make such measurements in galactically clear weather; the Milky Way is much wider around than it is thick, and it's chock-full of random blobs of gas and dust.

If you're an astronomer interested in blobs of gas and dust, then this is awesome. If you're a cosmologist, then this is beyond annoying. That space junk gets in the way of our pristine galactic images, making it difficult to find stuff *through* our Milky Way. This is especially true in the direction of our galactic core, the densest blob of gas and dust there is. It's a "zone" of the sky that we typically "avoid" because it's just too hard to look through.

It's a Zone of Avoidance.

And wouldn't you know it—the direction we're headed in, after you carefully subtract the motions of the Earth, sun, and Androky Way collision, is right *there*. Dang it.

We call it the Great Attractor, because who wouldn't?

The Milky Way, Andromeda, Triangulum, and a few dozen hanger-on dwarf galaxies form the Local Group, a gravitationally bound clump that's a handful of megaparsecs across.

The Local Group is a suburban town compared to the big city of the Virgo Cluster, more than 1300 galaxies crammed into a tight clump about a dozen megaparsecs down the road.

As I said before, clusters are the largest gravitationally bound objects in the universe, and you can probably guess where this is going: they're not done forming. Our Local Group is headed for the Virgo Cluster, leaving the country life for the glitz and glamour of the city.

But it doesn't stop there. Clusters themselves are gravitationally attracted to each other—they are massive balls of matter, after all—and are trying to form ever-larger structures. Unfortunately, definitions here get a little fuzzy. We know at the small end there are galaxies, groups, and clusters. And at the largest end is the cosmic web itself. In between is, well, stuff.

One name usually floated around is *supercluster*, which, just as it sounds, is something greater than a solo cluster. But despite finding them for decades, astronomers didn't really have a strict definition of one. If a structure was vaguely largish, it was a supercluster.

More recently we've come up with more robust definitions of these immense beasts. Since they're not gravitationally bound—they're not tied up into a tight little ball—we instead rely on the *motions* of galaxies to define them. Superclusters aren't yet done growing, so instead of looking at what they *are*, we must look at what they *will be*.

These new kinds of maps reveal the flows of our local universe—the agglomeration of structures into larger structures.[7]

Oh, and we've begun to map the universe within the Zone of Avoidance. While visible light gets snuffed out by our dusty galactic center, other wavelengths like radio and infrared sail right through, giving us the barest hint of the galaxies sitting on the far side of the Milky Way.

Combining the initial mappings within the Zone and a more respectable definition of a supercluster reveals what's going on with the Great Attractor. Remember how I said we live in a hierarchical universe? Yes, I'm repeating myself because it's that important.

Structures in our cosmos assemble themselves from smaller objects, and the Great Attractor is just the latest phase of that extragalactic construction project. The Local Group and a few other objects are collecting together toward the Virgo Cluster, which sits at the heart of the Virgo Supercluster. But the Virgo Supercluster is just a side branch of a much larger collection: the Laniakea Supercluster. And *that* supercluster has its own heart at the center: the Norma Cluster.

When you look in the direction of the Great Attractor, you're looking right at the Norma Cluster. But the Attractor isn't so much a *thing* as a *place*, the focal point of a process set in motion billions of years ago. Given enough time, every galaxy, group, and cluster within the Laniakea complex will collapse into a single uber-massive cluster.

Except it won't.

The cosmic web is not long for this universe. It's a transient, effervescent feature. A relic of the past, still persisting in the present but doomed in the future.

We live in a hierarchical universe, but we also live in an expanding universe. Structures like galaxies, groups, and clusters form *despite* this expansion. Although every galaxy is moving away from every other galaxy, this is only on average. Tiny instabilities in the early universe grow by gravity and, despite it being the wimpiest of the forces of nature, give it enough time and the pull of Newton and Einstein will do its indomitable work, growing structures bit by gaseous bit.

But the expansion is inevitable, and as I'm about to tell you in the next chapter, it's getting worse. Galaxies are still in motion, set off by gravitational interactions initiated in the distant past, but that motion is slowly grinding to a halt. Five billion years ago, the engines of creation shut down.

The Local Group will never reach the Virgo Cluster. The Virgo Supercluster will never reach Norma. Laniakea will never fully condense, and eventually it will be ripped apart.

In a few tens of billions of years, the cosmic web, with its beautiful, intricate lacework of filaments, walls, and knots—the largest pattern found in nature—will be gone.

All gone.

CHAPTER 11

THE RISE OF DARK ENERGY

When the universe was about nine billion years old, on the eve of the formation of our own solar system, after spending ages in calm complacency slowly spinning the cosmic web from loose threads of galaxies and dark matter, a crisis erupted as a hidden force began to wrest control of cosmological fate from the hands of gravity.

In the 1990s on Earth, the geopolitical cold war had finally ended with the collapse of the Soviet Union, but growing tensions within the astronomical community were now running hot. The stakes were never so high. The ultimate future of the universe hung in the balance. Well, to be fair, the universe is going to do whatever the universe was going to do anyway, but we'd like to understand it before it actually, you know, happens.

Cosmology, the science of the universe, is a game of taking almost stupidly simple questions and asking them to the whole entire cosmos as a single physical object. But, as we've seen, simple doesn't mean easy, and straightforward questions like "Hey, how much does that galaxy over there weigh?" can turn out to have some frustratingly surprising answers. So astronomers, in fine academic form, instead of taking the difficulties of measuring the mass of objects like galaxies and clusters as a proper warning from nature to back off, ramped up the observations and attempted the same feat on cosmological scales. Because *science*.

And immediately they ran into problems.

To explain the problems, I need to talk about geometry. It turns out Kepler was sort of right hundreds of years ago—the universe is ruled by geometry that can predict the future of the universe, just not *divine* geometry that can predict who my true love is. So, half right. Not so bad.

The language of the cosmos at the very largest scales is general relativity, and as I'm totally sure you remember from a few chapters ago, general relativity connects *stuff* (matter, energy, etc.) to *geometry* (the bending, warping, and flexing of space-time). This language can be applied to small systems like planets orbiting stars and black holes, and it can be applied to large systems, like the whole flipping universe. Seriously, like Einstein himself did shortly after inventing the concept. The equations of general relativity aren't picky (insert "general" pun here); to solve cosmological problems, the recipe is simple.

Tally up all the matter and energy contents of the universe. Be sure to include any nonvisible matter, radiation, and neutrinos. Everything. Stuff that into one side of Einstein's equations. Solve said equations (this is the messy part, but lucky for you, we get to skip over it in this narrative). General relativity then tells you how the whole entire universe bends, warps, and flexes.

Now what? Well, again with your handy general relativity tool kit, you take that knowledge of bending, warping, and flexing, and you can predict how it will eventually evolve. Voilà: the future fate of the universe in your hot little hands. We know that the universe is expanding, thanks to Hubble, but just how fast is it expanding? Is it speeding up or slowing down? Will it stop and turn around? Will it coast to a stop? If we fast-forward a hundred billion years from now, what will the universe look like?

The answers to these questions depend on how much of what kind of stuff the universe contains. Heaps of matter will drive a different expansion rate than heaps of radiation, and both of those will certainly result in a different fate than just a thin sprinkling of matter. So to get some clarity on the issue, it's absolutely essential to get a complete and total census of the material population of the cosmos.[1]

But wait, there's more. You can flip the equations of general relativity around. If you know—or can measure—the geometry of the universe, you turn the relativistic crank and learn the contents. All it takes is being able to measure the geometry of the universe, which, as you might imagine, could be a little challenging. Go ahead and make a bet now on which is harder: measuring the contents or measuring the geometry.

When I say geometry, I don't just mean circles and triangles; the geometry you learned in high school and promptly forgot until you had to paint your living room is only a subset of the full picture. That's the realm of what's called Euclidean geometry, which as you might have guessed was developed by Euclid himself, more than two thousand years ago. And that picture is all about flatness: circles, triangles, angles, lines, and everything all living in a plain, flat universe.

But general relativity is all about curvature—so, step number one, how do you define "curvature"? You have an intuition that the Earth's surface is curved, but your own backyard is relatively flat. How can we define curvature in such a way that we can apply it to any system that we care about, and specifically to the system that we care about the most: the universe?

Here's one way: parallel lines. If you draw two parallel lines on a piece of paper, and extend those lines to infinity, the lines will stay parallel. That's kind of the definition of "parallel," so I hope that's not a big issue for you. But try drawing some parallel lines on a globe. Start at the equator with two tiny lines, perfectly parallel. Advance each line northward in a perfectly straightforward way, never turning left or right. Soon enough, despite your best efforts, the lines will intersect—at the north pole.

You guessed it: the globe, and the Earth, is curved.

Another method is triangles. If you draw a triangle on a flat piece of paper, the interior angles will add up to 180 degrees. That's also the definition of a triangle, so we should be good. Now draw a triangle on a globe. Add up the interior angles. You'll get a number larger than 180, I guarantee it. Don't trust me? Do it yourself, right now. I dare you.

If you happen to have a horse saddle lying around (I won't judge), you can play the same games, but you'll find the opposite result: initially, parallel lines will spread farther apart, and triangle angles will fall short of 180 degrees.

The beauty of these definitions of curvature is that they can be applied to any number of dimensions—most important for us, three: the spatial dimensions of our universe.

That's great! Easy-peasy, we just need to bust out the markers and see how triangles and parallel lines on billion-light-year scales behave. Thankfully, nature provided the measuring tools already: the cosmic microwave background. We know, based on our knowledge of the nuclear realm, about the sizes and scales of the minute bumps and wiggles in that afterglow light pattern—we know how big they were when they were formed. And we know how big they are now, eons later, because they're right there, projected onto our sky.

Beams of light make for excellent cosmic Sharpies. Two beams of light, initially parallel and given enough distance, will eventually trace out the geometry of the universe on the grandest of scales. Sure, little things like galaxies (and yes, in a cosmological sense entire galaxies are considered "little") will distort their paths, but we want to paint a much bigger picture. In the intervening billions of years between generation and acquisition of those light beams, they will either remain parallel, spread apart, or converge. By comparing what we know about the typical sizes of patches in the cosmic microwave background to what we actually measure, we can literally measure the geometry of the entire universe.

It's flat.

Like a pancake. Like Kansas. Like a board. The universe is flat. Parallel lines stay parallel over the course of billions of years and billions of light-years. Our universe appears totally geometrically flat to a ridiculously small margin of error—we're talking one part in a million here, folks.[2]

When this answer—that we live in a flat universe—first started to become apparent in the 1990s with the first high-resolution maps of the cosmic microwave background, it was totally fine for the theorists. Why, they had been crowing about a flat universe for decades now, and they were glad the observers were finally catching up. Their reasoning came from inflation, that poorly understood but seemingly necessary process in the very early universe, when the entire cosmos blew up many orders of magnitude in the blink of an eye.

The process of inflation sent the true scale of the universe far, far, beyond our relatively pathetically small observable bubble. The universe could have any curvature it liked, like the surface of the Earth or a saddle, but it wouldn't matter. Our observable patch is so small that we're essentially guaranteed to measure a flat universe. Just like the Earth is round, but your backyard is so small it appears flat (insert usual caveats about small-scale deviations like groundhog holes not mattering for our measurements of the bigger picture).

Inflation went one better. To measure a certain degree of flatness today, with a giant universe, meant that the universe had to be even more flatter when it was smaller: the dilution of matter should have driven us far, far away from perfect flatness long ago. Inflation solves that problem by making the true universe so gigantic that we can't help but measure flatness. Nice.

If the inflation story is correct—and even though we don't know the characters, we understand the general plotline—then we *must* live in a (locally) flat universe. We don't have any other choice. So it's no surprise that the cosmic microwave background reveals precisely that answer.

But nobody else, most especially the astronomers, bought into that argument. Hence the *mild disagreements* of the 1990s.

The problem was that the measurement of flatness was simultaneously *also* a measurement of the total contents of the universe, due to all that general relativity stuff you just read about a minute ago. And the astronomers were already busy running around weighing all the galaxies and clusters they could train their telescopes on, and they were falling far short of the predicted number.

Like, less than a third. The total mass of everything in the universe, including even dark matter, was less than 30 percent of the matter necessary to match up with the observations of a flat cosmos.[3]

Hence, argument. The theorists accused the observers of being lazy and not working hard enough in their measurements, because surely they were missing something. The observers countered that theorists should take a break from the chalkboards for once and actually look at the universe around them that was abundantly not agreeing with their predictions—perhaps they forgot to carry the two in one of their calculations? And the cosmic background is pretty far away, even for astronomy. You sure you got it right?

You know, the usual stuff.

In comes a third approach, also tied to general relativity: the expansion history. Matter and energy tell space-time how to bend, and the bending of space-time tells matter how to move. So you can measure geometry, contents, *or* behavior to get at the underlying physics. And the behavior in this case is, of course, the expansion of the universe.

What Hubble managed to measure back in the early twentieth century was the *local* expansion rate, based on a relatively small sample of nearby galaxies. And in cosmology, "local" is synonymous with "today." The light from those galaxies didn't take *that* long to get here—it might as well have been from just last week, cosmologically speaking. So Hubble's measurement, and any other measurement based on close galaxies, is the current expansion rate of the universe *right now at this very moment.*

But the universe could have had different expansion rates in the past. Indeed, the entire concept of inflation depends on that ludicrous-speed expansion in its early moments. And with radiation and then matter taking center stage at different times with different densities, that can affect how quickly (or, for that matter, slowly) the universe expands.

To get these measurements we have to go deep into the universe, pushing further with surveys and observations than we ever have before. Redshift by redshift, galaxy by galaxy, we need to reconstruct the expansion history of the universe. Of course there will be limitations—the dark ages won't have much to offer us, at least not yet—but galaxies arrived on the scene pretty early in cosmic history. Surely there's some method to capture their distance and velocity, and to combine that with the data from as many other galaxies as possible.

That's the key; that will resolve the tension, one way or another, between the camps of scientists who have spent decades refining their techniques and sharpening their arguments but haven't come to a solution. It's the Kepler-Brahe and Shapley-Curtis debates all over again, centuries later.

To reach tall heights you need a ladder, and to pierce deeper into the celestial realm than our ancestors would have ever dreamed possible, you need . . .

. . . a cosmic ladder. No, that's not a joke. I mean, I totally set it up as a joke, but "cosmic distance ladder" is a real-life phrase used by real-life scientists to refer to a real-life concept. The challenge with measuring truly astronomical distances is that objects can be very far away *even for an astronomer*.

We've already met parallax, which after centuries of frustration finally led to an accurate distance to another star. Less than a century later, Edwin Hubble used Cepheid variable stars to confirm the remoteness of Andromeda. If you recall (which, by the way, is an academic code phrase meaning "you should have recalled"), Hubble couldn't use parallax for his groundbreaking work because parallax just wasn't good enough: the distances were too great, and the back-and-forth wiggles were too small to measure.

But even Cepheids, as useful as they are, can only be used to capture distances to nearby galaxies. While that technique swaps out a very difficult measurement (a distance) for a relatively easier one (brightness variations), you still need enough raw light to get the job done. If the Cepheid is too far away, you simply don't have enough photons to work with, and you're back to being hopeless.

So we need something else, some other way to hook into the distance of an object without *actually* having to measure the distance. And preferably we need some of these to overlap with at least a few known Cepheids. We could use Cepheids to reach farther than parallax because we had a few of those variable stars near enough that we could practically break out the astronomical measuring tape and pin them down. Once the method was safely confirmed, we could extend the Cepheids into the unknown depths without losing too much sleep over the issue.

So parallax and Cepheids are the first two rungs of a ladder, where we (hopefully) use a series of different objects and techniques to (hopefully) take us deep into the cosmic depths, each more distant method overlapping with a closer, vetted technique. A ladder that lets us measure cosmic distances. A cosmic distance ladder. See, I told you it was a real thing in real life.

Say you have two light bulbs of identical make, manufacture, wattage, color, and so on. Keep one bulb next to you, and throw the other one as hard as you can—or better yet, to continue with the experiment, walk it across the room and gently place it down. Turn on both lights. Break out your brightness-o-meter (I know you have one). Measure the brightness of the bulb next to you. Measure the brightness of the bulb far away from you.

Hopefully the distant bulb will be dimmer, and if you think a little bit about the relationship between dimness and distance, you could reasonably argue that if we had a bunch of such light bulbs scattered around the universe, we could use their relative dimnesses to calculate distances.

Unfortunately, flying around placing light bulbs at strategic cosmic locations would render the whole distance-measurement game moot, so we need nature to manufacture some for us. Many folks back in the day, like Newton and Galileo, had hoped/assumed that stars were of equal *true* brightness, allowing them to be used like our light bulb analogy, but sadly, that didn't work out well for anybody.

If stars don't cut it, what can? Whatever this *standard candle* could be, it has to be (a) common, so we can get a lot of measurements, (b) very bright, so we can see it reliably from far away, and (c) actually standard, so we can compare the measured to the true brightness and get work done.

In almost all circumstances, nature is sneakily cruel to us, offering only tantalizing hints of the mysteries of the cosmos, usually not enough to slake our thirst. But when it comes to cosmological questions, we have caught a couple of lucky breaks (it's still up for debate if nature willingly intended this or if they were just lucky accidents that slipped by her usually vigilant gaze). One case was the cosmic microwave background, which confirmed the big bang picture and unlocked the modern age of cosmology, and the other has to do with supernovae.

Stars are born, stars live, and stars die. Most stars, like our sun, simply exhaust themselves and extinguish their nuclear flames with only a modest flare or two to signify the ends of their lives. Some stars, the big ones, mark the end of their short, furious lives with explosions that literally light up entire galaxies—the supernovae.

Astronomers had been sporadically spotting new stars for millennia, and it was Tycho Brahe who named them *novae* when he extensively wrote about one that appeared in his sky one evening. By the late 1800s, it was realized that some novae were much brighter than others. Superlatively so, and the word *supernova* came into fashion.

As is the usual custom in science, the more a phenomenon like migration patterns or beetle larvae is studied, the more divisions, classifications, and subgroups get attached. It's no different with (super)novae, which, thanks in part to the work of our dear friend Fritz Zwicky, are helpfully organized into several different classifications with unhelpful naming schemes.[4]

Honestly, we don't need to care about the naming schemes and their origins and definitions. I know three of you are now going to throw away this book in disgust because I'm not going into detail on the subject, but for the rest of us, we're just going to pay attention to one particular kind of supernova: the type Ia. We're going to care about it the most because that's the one that cosmologists care about the most, and this is a book on cosmology.

In the 1990s astronomers noted, probably because they were looking for something exactly like this, that type Ia supernovae have a few desirable properties. While these kinds of supernovae are relatively rare, popping up a handful of times per century per galaxy, there are so many galaxies in the universe that they're basically happening all the time. They are exceedingly bright too: for a few weeks, a single supernova detonation will release more energy than an entire galaxy. That's a few hundred billion stars, for those keeping score. This means that we can see type Ia supernovae all over the place, even from incredibly distant galaxies.

So they're common and they're bright. Two out of three criteria met—but what about the third? You know, the most important one: do they have the same brightness? No, they don't. Well, almost, kind of.

When a supernova goes off, it quickly reaches a peak brightness, then slowly fades down over the course of a few days or weeks. Different type Ia's have different peak brightnesses, and they also have different cooldown times, which at first blush isn't surprising: supernovae are going to do whatever they dang well feel like. At first this might look hopeless as a standard candle, but the ever-ingenious astronomers noticed a pattern: the brighter a supernova reached at its peak, the longer it took to cool off. And not in a general, vague, hand-wavey sort of way; there was a very clear mathematical relationship between those two quantities.

That was the key that opened the door to the distant universe. Type Ia supernovae aren't standard candles, but they are *standardizable* candles. With a little bit of finagling,[5] you can capture a random supernova in some remote stretch of the far-flung cosmos and calculate its true, inherent brightness, the same brightness you would measure if you were right there in front of it, having your face melted by the blast. You can then compare the true brightness to the brightness you measure safely on Earth, and calculate a distance.

Even at a distance of fifty million light-years, this type Ia supernova still manages to dazzle, briefly expending more energy than its entire host galaxy. (Image courtesy of NASA / ESA.)

But wait, there's more! The magnificent explosions are cacophonies of light, featuring the usual assortment of absorption and emission spectral lines that astronomers of the nineteenth century learned to love. A reliable spectral line means you can get a redshift, which means a velocity (and if we can't use the spectrum from the supernova itself, we can always use the one from its host galaxy). This is the exact same technique that was unlocked more than a hundred years prior, was utilized to great effect to discover the motion of stars, and decades later revealed the expanding universe to Hubble's amazed eyes. And now it was brought to its ultimate conclusion, its final form: studying distant dying stars to prize out one last datum of cosmological significance from the fading embers of those cataclysmic explosions. Their stellar sacrifice is our gain, thanks to a practice developed in the steam age.

Two key pieces of info for the price of one (difficult) measurement: an unlikely gift from nature. If a type Ia supernova goes off in a distant galaxy, you can capture not only its distance but its speed. So not only do you get to extend the cosmic distance ladder to new celestial heights, you can also get a handle on the expansion rate between us and a distant point. And by figuring out the expansion rate at different points in deep time, you can tease out the matter and energy contents of the cosmos and figure out why all the other astronomers are hating on each other.

Of course, there's way more to the distance ladder—it's more than three rungs. There are more steps in between Cepheids and supernovae with fantastic names like RR Lyrae variables and the Tully-Fisher relation. Those aren't as useful for cosmological purposes directly, but they do serve as important cross-checks: each rung on the ladder isn't entirely separate, but overlaps with others, so we have multiple independent ways of measuring and calibrating distances. For the craziness I'm about to share with you, it's important to mention that part, because what cosmologists have recently revealed about our universe is easy to dismiss (that's how weird it is) if you don't know how robust the measurement really is.

I'm also purposely keeping you in the dark (for now) about the actual physics of type Ia supernovae, because as I said with Cepheids, *it doesn't matter.* What matters for the ultimate goal—measuring the expansion history of the universe—is that we can identify a standard candle. Who cares how the candle is made? All we need to know is that we can estimate its true brightness to a certain level of accuracy. Let's not make our lives messier than need be. Words of wisdom from the cosmological community indeed.

So here's the game plan: measure a bunch of type Ia supernovae (only type Ia behave in this friendly way, so unfortunately we have to stick to them). Measure a bunch more. Toss a few more in for good statistics. Calculate their distance and speed. Estimate the expansion history of the universe over the past few billion years. Use general relativity to calculate what the matter and energy contents have been over those past few billion years. Resolve dispute among astronomers and cosmologists. Sit back and relax.

Two independent teams in the 1990s went about exactly that, and their goal was to measure the deceleration, however slight, of the universe's expansion. That's because *any* amount of matter, however slight, would eventually pull in the reins on our current expansionary phase. It may not be enough to fully stop it—that requires a lot more matter than we could ever find—but the presence of matter in the universe slows down expansion over the course of billions of years. So by carefully pinning down that deceleration, they could tell everybody else what's out there in the cosmos.

The two teams, working independently, managed to measure a deceleration—but with a minus sign in front. The distant supernovae that they collected were *too* dim. They were farther away than they ought to be. It was 1998, and after checking their work over and over again, and a few furtive phone calls between the two groups ("Are—are you seeing what we're seeing?"), the world came to know that the expansion of our universe is accelerating.[6]

The collaborations also quietly scribbled out the word "deceleration" from their project subtitles.

We live in an expanding universe; it's getting bigger and bigger every day. We've had a century to become grudgingly comfortable with the concept. But now, at this very moment, the expansion is accelerating. It's getting bigger and bigger *faster and faster* every day. To really drive the point home, the universe is expanding faster *than if it were totally empty of all matter*. Like finding your favorite comfort food at the buffet, there's no going back.

In a replay of the Shapley-Curtis debates from generations past, the supernova results were the ultimate compromise. They satisfied everybody by satisfying nobody. Everyone was right because they were all wrong. The answers clicked into place in a way that nobody enjoyed. You go into the doctor with a broken finger and a nosebleed, and you come out with your amputated finger sewn onto your nostril. You're *technically* healed, but not in the most pleasant or straightforward way.

The universe is indeed flat—parallel lines stay parallel, and triangles add up to an agreeable 180 degrees. The inflation theorists and cosmic microwave background observers were vindicated. *And* the universe is only about 30 percent matter. The astronomers had done a fine job accounting for all the mass after all. What made up the difference, what filled up the universe from edge to edge, was a previously unknown substance (or, at least, an effect). This substance is causing the expansion of the universe to accelerate unbridled.

We had a good thing going with dark matter, sounding all cool and mysterious and a little sassy, so why not extend the concept? We have *absolutely no clue* what's behind this accelerated expansion, so let's call it—drumroll, please—dark energy.

Welcome to the modern universe. It doesn't make any sense.

What is dark energy? There isn't much to say on the subject, because if you didn't catch it the first time, we have no idea what it is. "Dark energy" is a sweet name for an observed phenomenon—the accelerated expansion of the universe—but that doesn't really illuminate (snicker) the *cause* of that expansion. It's worse than our current problems with dark matter; at least there we have some ways of navigating the theories and putting experiments online. With dark energy, even twenty-plus years after the initial detection, we're still in the groping-around phase.

But let's paint the picture, hazy as it is, as we've got it right now. We can start with the reason we call it "energy," and it's not because "matter" was already taken and there are only two kinds of things in the universe (and really, if you want to get pedantic—and who doesn't?—matter and energy are two sides of the same coin, but that's another book).

It's your birthday and I get you a present, because I'm pretending to be your friend for the purposes of this explanation. You undo the intricate bow and carefully unwrap the package, opening the box to find . . . nothing. Completely empty. A pure vacuum, in fact. No matter, no radiation. Not a single particle, not a single photon. Absolutely pure nothing.

You politely mask your disappointment and thank me, internally rolling your eyes that I'm using a special event like your birthday for another stupid science thought experiment. You sigh, preparing yourself for the inevitable monologue. Here it comes.

Your empty box is anything but—it's actually full of dark energy. Dark energy fills the cosmos to the brim. It's a property of the vacuum of space itself. Have a small empty box? You have a little bit of dark energy. Big empty box? A lot more dark energy. Since we live in an expanding universe, we're getting more and more vacuum, more empty space, and hence more dark energy every day.

Now it just so happens that when you take a material with this property—that you get more of it when you expand the volume of its container—the math of general relativity produces a surprising result, and you can probably guess what it is: you get accelerated expansion. The very fact that dark energy is a property of the vacuum means that it drives the continued creation of itself. The universe has dark energy. It expands. It has more dark energy. This pushes the expansion a little bit faster. Even bigger universe. Even more dark energy. Even more accelerated expansion. Rinse and repeat if desired.

And before you interrupt me, I'll interrupt myself: you may be familiar with the concept of conservation of energy (usually incanted in some fashion like "Energy cannot be created or destroyed, it can only change forms"), and you're probably asking yourself why this first pass at explaining dark energy doesn't violate rule 1 of How Things Work. The answer is very simple and very annoying: energy isn't always conserved. It's conserved in a few special systems, especially systems found in homework problems of physics textbooks, but nature is much more subtle than that. In a *changing* universe (like, say, one that's expanding), energy can be added at will. It's just not a big deal, and we're all going to have to get comfortable with that if we're going to make any progress, OK?

Anyway, this accelerated expansion business didn't get started in earnest until about five billion years ago. It's not like dark energy just *poofed* into existence by some poorly understood process. *Dark energy has always been here*. It's a property of the vacuum of space-time. It's there, right in front of your face and inside your very bones. It's been with the universe since the earliest moments of the big bang itself. But it's been hidden, in the background, unimportant.

It's a game of densities: the same game that's been played for billions of years in our expanding universe. Matter finally won out over radiation, eventually causing the release of the cosmic microwave background, because the expansion diluted the radiation and dropped its density well below thresholds where anybody would notice or care. And while matter had its eventual—and inevitable—triumph, its reign was not forever. There's only a fixed amount of particles—dark or otherwise—available in the universe. Day by day, cubic meter by cubic meter, the density of matter has been dropping.

But dark energy's distinguishing feature is *constant* density. The bigger your universe, the more total dark energy there is. While matter is dominating, its gravitational attraction overwhelms any inclination dark energy might have to accelerate the cosmos. Indeed, the expansion of the universe slowed down over the course of the first few billion years. But like too little butter spread over too much bread, matter lost its grip on the fate of the cosmos. It's simply not a player anymore. Starting five billion years ago, the density of matter slipped below that of dark energy and continued falling. Dark energy dominates the modern-day universe. It's the single largest component. It's in the driver's seat now, and it doesn't even know where the brake pedal is.

I think it's important to remind you that you can sleep soundly tonight—dark energy only makes itself noticeable on large scales. Clusters are still gravitationally bound together. Galaxies safely keep their stars within their embrace. The planets still twirl in their waltzes around their suns. Dark energy is persistent but weak—any place where another force is stronger, the accelerated expansion can't take hold. Just as an expanding universe doesn't mean the Earth is expanding, an accelerated expansion only takes place in cosmological settings.

Unfortunately for the cosmic web, it can't sleep soundly tonight. Beyond the scale of already-formed clusters, there's not enough attractional oomph to overwhelm the tearing tendency of dark energy. Like a swimmer caught in the riptide, getting farther away from the shores no matter how hard she fights.

It's happening in the voids first. They're already empty of most matter, and it's there that dark energy is most dominant, expanding and inflating them, pushing their boundaries ever larger. The tenuous cosmic web exists now only as gossamer strands and almost-transparent walls between the indomitable nothingness of the voids. Eventually those grand structures will dissolve. As the cosmic web becomes unspun, only the isolated clusters, mere pockets of gravitational attraction strong enough to stand fast, will remain, becoming remote islands separated by vast black wastelands ruled by one thing and one thing only: dark energy.

Pretty dark stuff (pun most definitely intended). I'll get to the future fate of the cosmos in more gruesome detail later, but for now I want to clear up a few loose ends. First off, this business of energy. Hmmm, what else fills up the vacuum, permeating all of space-time within the entire universe? Good question—I'm glad you asked. Quantum fields do! The buzzing and vibrating substrate has all the needed properties to explain dark energy. And it is an *energy* too—a vacuum energy. At the ground state, the bare minimum allowable configuration, the quantum fields that constitute our reality have some base energy, and they are perhaps ultimately responsible for the accelerated expansion of the universe itself.

We've broken open the quantum field In Case of Emergencies box before, to explain the driving force behind inflation and to seed the growth of the largest structures in the universe, so at first blush, this seems like a natural extension of those strategies. Have a mystery in the universe? Just blame quantum fields—look, look, they're practically *oozing* guilt on their faces. It's easy. Everyone will believe you.

One small, tiny, niggling caveat: it doesn't work. I have a confession to make. When physicists go to calculate the actual amount of vacuum energy— like, actually produce a number instead of just talking about it vaguely—they get infinity. That's right: infinity. It turns out that all the wiggling and jiggling those quantum fields do at a subatomic level keeps adding and adding to itself without end. The fundamental (ha!) problem is that the wiggles can be as small as they want. It's like trying to listen to an orchestra with an infinite number of instruments that can play at every frequency imaginable. All the sounds add up to an infinite amount of energy and your ears bleed and your head explodes.

That's . . . kind of a problem, which we saw earlier in the saga of physicists first trying to extend the work of Dirac and marry special relativity to quantum mechanics. Fortunately for most calculations of importance, the absolute value of this vacuum energy doesn't matter. I know it's weird and nonintuitive to think about, and I'm truly sorry about that, but that's physics for you. You can do your particle science on the first floor or the tenth floor or the infiniteenth floor; it doesn't matter.

Except when it matters, as in the case with dark energy. We know very well (by now, at least) how much vacuum energy, if it is the culprit here, is driving the accelerated expansion. It's a very tiny and definitely not infinite number: somewhere in the ballpark of 10^{-30} grams in every cubic centimeter of the cosmos—ten measly hydrogen atoms per cubic meter, give or take, spread across the entire universe. That's all it takes to give us the spectacular and stupefying accelerating-cosmos results.

When we attempt a naïve prediction of this number based on quantum field theory by, say, only adding up frequencies to some reasonable threshold like, I dunno, the Planck scale or whatever, out pops a number clocking in at 10^{90} grams per cubic centimeter. That's wrong. Very wrong. Deeply, uncomfortably, you're-obviously-missing-something-fundamental-about-the-universe wrong.

So maybe quantum fields aren't the best path forward to explaining dark energy. But even if we can't get the number right (or even *close to kind of* right), dark energy still *acts* as if it were a vacuum energy. Even if the true source is something else, accelerated expansion has the behavior—just unfortunately not the magnitude—of space-time-filling quantum fields permeating all of the cosmos.

Basically, we're stuck.

There's another way to frame this problem in general relativity, which might or might not be helpful, depending on your level of pessimism. This framing can be sourced directly to old Einstein himself. Remember how, when he first manufactured relativity in the general sense, he didn't know about the expanding universe? And how the equations gave him the flexibility to toss in an extra constant to maintain a static cosmos? And then as Hubble astonished the world, he scrubbed out the so-called cosmological constant when nobody was looking?

Well, here we are again, finding ourselves in a situation where it looks like we need to add that constant back in after all. Not to maintain a static universe against any movement, but to power the accelerated expansion. The reason this might help is that it's an alternative way to formulate the conundrum, which might lead somewhere promising, since a cosmological constant sits on the "curved space-time" side of the relativity fence. The downside is that a cosmological constant is perfectly identical to a vacuum energy, which sits on the "matter and energy" side of the equations of general relativity. So you haven't actually moved anywhere.

Still, could dark energy really be a fault of gravity rather than some vague, poorly understood part of our universe? It's the same question we had with dark matter, and it has the same problems. As nauseating as vacuum energy is, there appears to be (at the time of this writing) no modification and/or extension of general relativity that competently explains dark energy.[7]

Part of the problem is that we're in the dark (ha!) not just theoretically but observationally. It's been a couple of decades since dark energy first burst onto the scene in scores of supernova detonations seen across the universe. Astronomers are not idle folks, and in those years they've collected heaps of evidence and come to the conclusion that yup, it's still there.

Take, for instance, those baryon acoustic oscillations, the fancy-pants term for the rockin' sound waves in the early universe. When recombination smoothed the troubled waters of the young cosmos, those waves got frozen in, leading to a slight—but detectable—impression in the arrangement of galaxies on very (very) large scales. That impression acts as a standard *ruler*, as if you tossed a bunch of yardsticks (metersticks in the international print of this book) around the cosmos. Since we know the size of the oscillations, we can compare that to the size we measure in the sky, and reconstruct the expansion history at those points.

It's a completely independent way of targeting dark energy, and it reveals . . . dark energy. Indeed, if dark energy hadn't happened to our universe, galaxies would have continued their cosmological building program, erecting ever-larger structures, and that process would have washed out any primordial imprints altogether. The very fact that we can still detect that faint impression leads us to conclude that dark energy is happening.

Despite decades of our best observational efforts, we've only been able to, at best, confirm the existence of dark energy, without actually learning a lot more about it. Has it changed with time, or is it truly a cosmological constant? Is it connected at all to dark matter? Are there multiple sources of dark energy? If it's a vacuum energy, why can't we get the numbers to come out right? Is dark energy related to inflation, which was driven by another (mysterious) quantum field, and if so, why are these effects separated by billions of years and orders of magnitude in energy?

All questions, no answers.

Our first dart thrown at the dark energy board is miles off, but we don't have any other decent (even half-decent; heck, we'd even accept 1 percent decent at this stage) ideas.

The modern picture of our universe, as painted by hundreds of observations and experiments independently searching for answers and cross-checking each other, from the cosmic microwave background to distant supernovae to the weights of clusters of galaxies, is cold and bleak: 13.8 billion years old, composed of less than 5 percent normal (light-loving) matter. One-quarter dark matter and three-quarters dark energy. It's geometrically flat, but the expansion is accelerating, for reasons we don't fathom.

We call it "concordance cosmology," as it's the result of many different lines of research that all point to the same bleak conclusions.[8] It's a completely different universe from the one explored hundreds of years ago. That universe was complicated and messy, but small and hot. Cozy and alive. The universe revealed in the modern age is old and slow, well past its prime and dominated by mysteries piled on mysteries. It turns out that the efforts of generations of scientists over the course of centuries have barely even scraped the cosmological surface.

CHAPTER 12

THE STELLIFEROUS ERA

But the universe is not dead yet, and we have unfinished business.

Against that backdrop of an old, cold, dark, expanding cosmos, fires still burn. Normal matter may make up less than 5 percent of the contents of the universe, and indeed you could erase out all the baryons in existence and the long-term history and fate of the universe would largely go on unchanged, but baryons do deserve some special discussion. After all, even with the development of neutrino and gravitational wave astronomy, stars and light are our primary views into the celestial realm. That was true hundreds of years ago at the birth of modern cosmology, and sometimes old tricks are still good; hundreds of years from now, I'm willing to bet that the good old-fashioned optical telescope will be at the forefront of astronomical research.

It may feature a mirror the size of small planet, but the basic gist will still be the same.

Four hundred years ago, Galileo Galilei revolutionized our understanding of—and our place in—the universe using a simple, small telescope. Three hundred years later, Edwin Hubble accomplished the same feat using a much larger version of the same instrument. The thirty decades between them saw an almost-annual revision of the cosmic cast of characters as new vistas were opened, new maps were charted, and new mysteries developed.

By the late nineteenth century, two major questions began to crystallize: how do stars work, and how big is the universe? We've been following the second question for the past few chapters, which at first had some surprising but respectable answers ("larger than you thought") but quickly led to discoveries that made it seem like nature was just playing a cruel joke on us ("by the way, your kind of matter doesn't matter").

Answers to the latter question were gradually taken up by a fervent group of astronomers and physicists who, over time, eventually gained self-awareness and named themselves *cosmologists*, studiers of the universe itself. At first a somewhat fringe and hand-wavey discipline, known for its startlingly inaccurate measurements, the field grew into respectability, and even popularity, with the discovery of the cosmic microwave background and large-scale maps of the cosmos.

But the traditional astronomers weren't asleep at the aperture in the twentieth century. Through careful (scientific, even) observations, scientists around the world and through the decades unraveled the mysteries behind stars and galaxies. Don't get me wrong—there are still about a million things we don't understand about the baryonic, light-loving world, but we've come a long way.

When we last left the nineteenth century, we turned to matters of cosmological interest. But there were still so many puzzles. So many varieties of colors and sizes of stars, in all sorts of groupings and collections, some sprinkled evenly throughout the galaxy, others clumped together. Some were isolated loners; others, complicated pairings, triplets, or more. New stars would appear, burning fiercely for days or weeks before fading back into obscurity. Some would pulse rhythmically over the course of months—or minutes!—without hesitation or interruption.

And then there were the nebulae—thin, wispy veils of dust and gas, sometimes associated with stars and sometimes alone. Some kinds of nebulae were later violently reclassified as entire galaxies, with creative classification schemes rapidly applied to the new order of celestial objects. Others maintained their original cloudy moniker but still held their secrets well.

Deeper probes sketched the shape of our own Milky Way as a thin disk, with twisting spiral arms and a bulging egg-yolk center, orbited by satellite galaxies and clumps of red, dead stars. More observations, especially with the opening of X-ray and radio astronomy, revealed even stranger creatures like pulsars and quasars, some within and some without our home galaxy.

The more nuclear fires were collected and categorized, the more the intellectual fires burned in the hearts of astronomers around the world: how does it all work, and are we connected, even in the slightest, to the celestial realm?

The revealed answer is frustratingly scientific: no, but also yes, in a technical sense.

Let's get the "no" part over with: the stars are almost incomprehensibly distant from us and do not affect us, here on the surface of the Earth, in any remotely conceivable way. Sorry, Kepler. There's no divine order to it all, and the position of a particular star or planet at the moment of our births does not have anything to do with anything. The forces that govern the motions of celestial objects are complex and chaotic. The regular patterns in the night sky are a coincidental effect of the Earth's rotation and revolution, not caused by anything fundamental in their nature.

Even the planets, close enough to be considered "ours," are thoroughly remote and isolated balls of gas and rock. Sure, *technically* gravity's influence is infinite: right now, as you read, you're feeling a slight tugging from the massive gravity of Jupiter, the largest of the solar system planets. But even that mighty giant, with its 317 Earths' worth of mass, is so far away that the gravitational attraction of this very book has more influence on you.

And the stars themselves? You can imagine experiencing our sun up close, its surface a roiling inferno, a cathedral of plasma and radiation. From the Earth, a serene yellow ball. The same ferocity placed light-years away? A pin-prick of light, a literal point in the night sky, without any dimension, easily overwhelmed by nothing more than a streetlamp.

But in a strange twist of fate, those distant stars are more than just tiny points of light sometimes visible in a dark enough sky. They're our cosmic cousins.

Given the winding and interconnected nature of scientific research, it's hard to pick a singular watershed moment when we first started to make the big connections that signaled a change in how we view the world. But for the sake of convenience and narrative simplicity, which I'm sure you'll appreciate, I'll start with two fellas named Ejnar Hertzsprung and Henry Russell and their handy little diagram of stars.

Before 1910, stars were just stars. Astronomers were used to the incredible variety on display in the heavens but had gotten little further than simply naming them. Red giants, white dwarfs, blue supergiants, and so on weren't the most creative or romantic names, but they served their functional purposes well. But how were these stars of different sizes and colors connected together? Complicating this connection was the fact that stars also had different levels of brightness, sometimes due to their very nature and sometimes due to their different distances from the Earth. If I didn't tell you how far away I was, you couldn't tell me how intrinsically bright my flashlight was, and vice versa.

Ejnar and Henry (not a crime-fighting duo, as their names might suggest) took a stab at boiling down all this mess into its essential essences and trying to discern what really mattered when it came to stars. Working with catalogs of thousands of stars, they applied a variety of methods to figure out their distances and hence their true brightness. Plotting those thousands of stars on a diagram comparing the brightness to the color revealed a puzzling pattern. Perhaps the most puzzling part of that pattern was the peculiar fact that there was a pattern at all.[1] Stars didn't just pick a random number from the brightness and color lotteries at the moment of their birth—almost all stars lived along a narrow diagonal strip, with brighter stars shifting to bluer colors. The largest stars formed a horizontal branch above that strip, and the white dwarfs were isolated in their own little island.

Fantastic, a pattern! Drat, another mystery. *C'est la vie scientifique.* This wasn't a solution to the problem of stars but, at least, a major, major clue. Which is a good enough start, I suppose. If I were Arthur Conan Doyle, I'd probably have my character say something really clever right now.

It's always amusing to read about a theory that everybody at the time figured was probably wrong, but nobody had any better ideas. In this case, the original thought was that perhaps stars start big, red, and fat, then slowly contract over their lives. The steady gravitational collapse provides a source of energy, powering their radiative expenditures.

Slight hitch: this process can make a star like the sun burn for ten or twenty million years, tops, and the biologists and geologists at the time were already pointing out that the Earth was a billion years old at the low end.[2]

The solution came a decade later, first suggested by (Sir) Arthur Eddington—of "let's go on a safari to test Einstein's relativity" fame[3]—once we figured out the whole nuclear physics and quantum mechanics and mass-equals-energy business happening at the subatomic level. It's a complicated story (of course), but let's take a high-level stroll.

The sun is a giant ball more than one hundred times wider than the Earth (in fact, there are boiling pustules on its surface larger than our entire planet), clocking in a whopping 333,000 Earth-masses of mostly hydrogen, a good fraction of helium, and a sprinkling of some heavier junk. Imagine yourself sitting in the center of that behemoth. It's just a little bit intense: the hydrogen plasma soup is crammed so tightly it's 150 times denser than water, the pressure is a headache-inducing 250 *billion* times greater than at sea level, and the temperature is a scorching 15.6 million Kelvin. I know, I know, an August day in Ohio can *feel* like this, but it's nothing compared to the sun's core.

With the hydrogen atoms crammed so tightly together, they overcome their natural electric hatred for each other and—replacing descriptions of complicated nuclear chain reactions with the simple word "fuse"—fuse into helium. Wonderful, the intensity of the sun's core is cooking hydrogen into helium; what does that get us? It gets us liberated energy, that's what. That's because the inputs into the reaction (four individual protons) are slightly more massive than the outputs (two protons and two neutrons glued together into a helium nucleus).

It's the *gluing* that's doing the work to provide the difference. Look at it this way: it takes energy to get your hands in there and rip apart a helium nucleus, so the formula is "helium + energy = separate parts," which can handily be rearranged as "helium − separate parts = energy." That difference in energy is precisely a difference in mass, just like Einstein taught us. So speaking of these reactions in terms of mass differences or energy differences is all the same, because mass and energy are equivalent, and the sun can glue together protons and spit out energy in the form of radiating photons.

There are also some extra products from the reactions, like positrons and neutrinos, and it's the detection of the neutrinos that lets us peek into the core itself and verify that yes, the fusion party is still raging deep inside the sun.[4]

This is a very efficient process, trading a little bit of mass for a lot of energy, enabling the sun to burn for billions of years and also enabling the biologists to say they told us so. Fine, we'll give them that one. To speak specifically for once, at the end of each nuclear chain reaction, the end products are 0.7 percent less massive than when they started, giving 26.73 million electron volts (hey, remember those?) of energy. Every single second, the sun chews through six hundred million tons of hydrogen, giving an energy production rate that exceeds a staggering 10^{26} watts. Yes, there's an awesome word for it: one hundred *yottawatts*. That's a lotta watts.

Of that incredible photonic output, most gets dumped into empty, useless space, and a measly 0.000000045 percent, or about one part per ten billion, actually strikes the Earth, half of which makes it to the surface during the day, providing the ultimate power source for all life on this planet.

Once we cracked the nuclear code, the Hertzsprung-Russell diagram fell into place, and it turns out we had stellar evolution completely backward: as stars age, they grow larger and more luminous. Who would have guessed? As a star consumes hydrogen, lumps of inert helium ash grow like tumors in its core. To compensate for this and keep the fusion party going, the temperature of the core rises, increasing the fusion rate and hence the size and luminosity of a star. Have a few duds join your party? Just crank up the music to compensate—that'll do the trick.

Even the dinosaurs, which roamed the Earth only a scant sixty-five million years ago—a blink of a cosmic eye, so to speak—knew a slightly smaller, slightly weaker sun. But this change is only relatively slight, and through the bulk of a star's lifetime, it lies on the central strip of the H-R diagram known as the "main sequence." This longevity is provided by hydrostatic equilibrium, the same kind of force-balancing exercise practiced with gusto by galaxy clusters. In the case of stars, the outward explosion from the nuclear inferno is equalized by the inward gravitational pressure.

But eventually (where "eventually" can mean anywhere from ten million years for the biggest stars to ten billion years for the sun to trillions of years for the smallest stars), the helium detritus grows too massive and forms an unmotivated core in the center of the star, forcing the hydrogen fusion party out of the house and into the front yard. The expansion of the fusion shell pushes the outer atmosphere of the star well beyond its normal limits. Separated from the nuclear core by a larger distance, the surface now cools, and the star branches off the main sequence, becoming a red giant. (Because it's red and giant. For once, astronomers named something appropriately.)

From here, different stars have different endgames, though they're all miserable in their own unique, special ways. For stars like our sun, the interior pressures and temperatures eventually reach such incredible heights (let's say one hundred million Kelvin for good measure) that helium burning begins. But helium fusion, leading to carbon and oxygen, isn't as efficient as hydrogen burning, so to stabilize against the inevitable collapse of gravity, the reactions must take place at a feverish rate, and the helium-themed party doesn't last nearly as long. In only a few minutes, a good chunk of the helium is burned off in a flash, which inflates the core and immediately cools it, temporarily shutting off fusion and leading to, over the course of ten thousand years, the collapse of the red giant.

But once the star settles down again, the fusion game picks back up, this time with a solid core of carbon and oxygen, surrounded by a shell of helium fusion, with *that* surrounded by good old-fashioned hydrogen burning. The party is now so intense it's in the streets, and it can only keep it up for a hundred million years or so before it begins climbing back up to red giant status. What follows is a series of violent seizures as different layers of the star abort and restart fusion in jagged procession. The expansion of the star and the intermittent violence in the core vomit the outer layers of its own atmosphere out into the system.

Once every hundred thousand years, a new spasm hurls more of the star's mass into surrounding space. The death throes are long, slow, and painful as the star tears itself apart, one vicious episode after another. Before long, the brilliantly hot core of unburnt carbon and oxygen, about the size of a respectable rocky planet, is exposed to space for the first time, searing the surrounding system with powerful X-rays.

But that period of violence leads to a moment of ephemeral and effervescent beauty. The remnants of the star are, by now, long since blown out in complicated patterns into the depths of space, and for a brief time, barely ten thousand years, the X-rays from the leftover core illuminate the decrepit remains, energizing the gas like a neon sign, with any unique elements giving off their distinctive spectral glows. While the core (now encumbered with the name *white dwarf*) remains hot enough to emit X-rays, the newly spun *planetary nebula* glitters like an ornament, a masterpiece of interstellar art, unique to that star, never seen before and never to be seen again. But when the core shuts down, the curtain finally closes, and stars like our sun leave the cosmic stage.

For stars larger than our sun, their end comes too, but much more quickly and with efficient mercy compared to the slow, drawn-out suffering of their lower-mass siblings. For them, their increased bulk can keep a gravitational lid on things, preventing runaway expansion and expulsion as a planetary nebula. Instead, the core grows hotter and hotter, reaching new levels of fusion intensity, with carbon and oxygen fusion kicking in at the billion-Kelvin mark, followed by silicon fusion at three billion Kelvin. Each stage is shorter and shorter as the fusion processes grow ever less efficient, the ending stages leaving the star looking like a nuclear seven-layer bean dip, layers going from hydrogen in the outermost layers to iron and nickel in the center.

It's a runaway process driven by continued, pounding gravitational collapse. The hydrogen burning lasts a few million years, followed by a million years of helium fusion. The carbon can power the star for only six hundred years (notice the lack of any other word after the number), with neon lasting barely a year. Oxygen takes a turn for a mere six minutes, ending with silicon forming a solid iron ball in the center of this massive beast in less than a day.

With iron, the cops finally come in to break up the scene. Fusing elements lighter than iron leads to energy gains as the binding glue increases. But above iron, it *takes* energy to form heavier elements. So the fusion process still happens, but there's no leftover energy to stop the crush of 10^{31} kilograms of material on top of it. The result: any errant electrons buzzing around get shoved into nearby protons, converting them into neutrons via the weak nuclear force. In a dozen minutes, the entire core of iron is transformed into a dense neutron sphere the mass of the sun and the size of your neighborhood.

Without overwhelming pressure, that's as compact as the neutron sphere can get. The superlative tons of atmosphere crush inward, meet resistance, and rebound outward.

As they say, boom.

These supernova explosions signal the deaths of the most massive stars in our universe, flinging newly fused material out into the surrounding medium. The energetics of the detonation itself, an instant that releases more energy than the entire lifetime output of a hundred suns, fuse a slew of elements heavier than iron, enriching the interstellar expanses with the rest of the periodic table.

These titanic outbursts, known as type II or core-collapse supernovae, aren't the only flare-ups in the cosmos. This is the common fate of massive, but isolated and lonely, stars. But most stars are found in double or triple (even septuple!) systems, and that leads to some other interesting situations worthy of note—mainly because they're hard to ignore.

A star's fate is largely set the day it's born. Its mass determines its future track on the H-R diagram. Low-mass stars lead long, but uneventful and not very bright, lives. Medium-mass stars like our sun burn for billions of years before turning inside out. Massive stars burn the nuclear candle at both ends, snuffing themselves out in millions of years. When stars are born in the same system, they will inevitably have different masses, because why should they be the same?

Thus the more massive sibling will go through its life faster, leading to the inevitable sad-sack white dwarf or neutron star. Its lower-mass relative, bound by gravitational chains, is forced to watch the whole ugly process play out. But it too eventually succumbs to a similar fate. As the star expands into its red giant phase and draws close, its outer envelope funnels onto its long-dead companion. When enough material accumulates on the surface, the increased pressures can ignite a nuclear burp—an intense but brief flash of energies, a nova.

196

In some rare cases, enough material can spill onto the surface of a companion white dwarf without small eruptions, saving its energies for a single insane outburst that triggers a chain reaction in the carbon-oxygen soup of the dead star itself, ripping it apart in an instant, ecstatic *whoosh*. Since white dwarfs always have about the same mass, the explosions have roughly the same brightness every time they go off. A standard candle, if you will; a powerful supernova of another kind—the type Ia—first recorded in detail by Tycho Brahe himself, and used hundreds of years later to map the expansion history of the universe and discover dark energy.

And you thought I would just leave you hanging about type Ia supernovae, didn't you?

Baryons may make up a small fraction of the contents of the universe, but they can pack quite a punch. But it's not all fusion factories and energetic explosions out there in the vastness of space. Stars may get all the attention, constantly strutting their radiative stuff and occasionally blowing themselves up, but there's another side to the baryonic coin in our universe—the nebulae.

Thin, wispy tendrils containing up to thousands of suns' worth of raw material. Mostly hydrogen and helium—no surprises there—but always sprinkled with traces of metals (in the joys of astronomical terminology, anything heavier than helium is called simply "a metal," because don't ask). With the naked eye, or telescope-enhanced eye, they generally come in two different colors: bluish and reddish. The red ones are typically near very bright stars, absorbing their high-energy output and spitting it back in more familiar colors. The blue ones hang out near less intense stars and simply reflect and scatter any incoming light, which, as in our own atmosphere, produces a warm blue tint.

While a cloud of gas and dust can hang out in the middle of nowhere for as long as it wants, occasionally it will receive a kick—say, from a nearby supernova or passing cloud—and when that happens, troubles set in. As the cloud pulls in on itself, pieces pinch off and catastrophically collapse, their internal densities and pressures climbing higher and higher. The more they collapse, the more they heat up, and the more they heat up, the more they radiate away that energy, cooling themselves and collapsing even further. It's the result when hydrostatic equilibrium is neither static nor in equilibrium.

The collapse continues unabated until, you guessed it, the pressures and temperatures reach that special critical threshold and hydrogen fusion ignites deep in the core—a star is born, surrounded by the beginnings of a planetary system. Once the star gets fully pumped up, it clears away its dusty nest, leaving behind any planets that may have formed from the leftover materials. The star lives its life, how long depending on how much gas got pumped into it during its nursing phase, and eventually dies, in one form or another, spewing its contents back to where they came.

That newly formed nebula, freshly ejected from the dying star, mingles back with the general interstellar milieu, finding new friends and new hangouts. The next generation of nebulae live their lives unmolested until a new instability crisis, repeating the pattern.

Stars. Nebulae. Stars. Nebulae. A cycle of cosmic birth, death, transformation, and rebirth, repeated since the first stellar denizens appeared on the scene all those billions of years ago. Each successive generation a little bit richer in elements, with more carbon, silicon, oxygen, iron, plutonium, potassium, and you know the rest of the periodic table.

After enough generations some of those stars could host planets composed of oxygen and silicon bound together to make a material known as "rock," with enough oxygen leftover to bond with hydrogen to make "water." Maybe some spare nitrogen gas to make "air." Add a healthy dose of sulfur, calcium, and phosphorus and a sprinkling of a few others, and you get "living creatures." Star stuff (or star poop, if you're feeling juvenile), formed in fusion factories in the hearts of stars or in cataclysmic explosions, regurgitated and recycled for eons. A process set in motion billions of years ago in a dark and lonely universe, hydrogen moved by simple gravity, reconstituted into bizarre and complex organisms capable of locomotion and cognition.

Sometimes those organisms even start asking questions.

Maybe Kepler was right after all, after a fashion. We're not communicating substantially with the celestial realm through any physical force, and the stars certainly don't tell us who to marry or the best time to wager on that horse. But the heavenly denizens did break themselves down to form the solar system, the sun and all the planets. So in a sense—a technically narrow but hauntingly beautiful sense, a sense that takes deep time and expanded horizons to realize—we are indeed connected to the stars, and they to us.

And beyond the scale of stars and nebulae tracing out their interconnected, sometimes-blowing-up lives lie the galaxies themselves. We've already taken the measure of the Milky Way, but we haven't cataloged its contents. It may not be a significant player on the cosmological scene, no more or less so than any other galaxy, but dang it, it's our home. We should be proud of it.

Galaxies got a major boost in the hearts and minds of astronomers about a hundred years ago, when Hubble firmly concluded that the strange, dusty spirals, at one time considered to be just another variety of nebula, were instead island universes—galaxies—in their own right, far removed from our own Milky Way by a vast expanse of vacuum.

But as our surveys grew broader and deeper, galaxies once again lost their importance, or at least their uniqueness. One speck of aggregated light isn't much different from another, as long as it faithfully traces out the underlying dark matter and gives us frames of reference for the occasional supernova. To a cosmologist, galaxies are a means to an end, a tool for measuring even greater things.

Still, that's not the whole story. There's more to a galaxy than meets the eye, and there's a heck of a lot that meets the eye. As we added more galaxies to our collections and mounted them in our display cases, we of course noticed all the peculiar differences. Many had the iconic spiral arms, quite a few were just general balls of gas and stars, and some were very lumpy and irregular. Of the spirals, some had dazzling pinwheels; some had just a couple of spokes; some had central bulges; and others had long, extended bars in their cores. Some galaxies were busting with star formation activity; some were long dead, dim, and red.

Besides the big ones (the biggest a few times larger than the Milky Way), the universe was littered with small dwarf galaxies. These little intergalactic rug rats averaged a bare 1 percent the size of our own galaxy. Sometimes they orbited larger galaxies, like the Magellanic Clouds, and sometimes they were isolated in the cold vastness. They too had a diversity of shapes: sporting spirals when they could, settling into elliptical monotony, or getting torn into irregular clumps.

Even more galaxies awaited the patient astronomer capable of deep observations at radio wavelengths: the blazars, the quasars, the LINERs, the Seyferts, the LIRGs. Fanciful acronyms and names used in place of real understanding. Some galaxies are quiet as a mouse, and some blast so loudly they can shine across the visible universe without breaking a sweat.

What's the common thread? Hubble himself attempted a classification scheme that, while not quite right, is still used because Hubble was awesome and I guess we respect the guy or something.[5] Once we figured out that the universe *evolves*, which itself was a major breakthrough, pieces of the galactic family puzzle began to click into place—though I should repeat my usual caveat that there's still a lot we don't understand.

There's something you should know about galaxies: they have secrets. Deep, dark secrets they hold close to their hearts. In fact, their hearts themselves are the secrets. They're infinitely black and all-consuming. As we've seen already, it's now understood that almost every galaxy hosts a massive black hole in its core. These black holes are tremendous beasts—at the *low* end, millions of times the mass of the sun.

Known appropriately as *supermassive* black holes, they are for the large part quiet. With no stars, dust, or gas nearby, they sit there, lurking in the shadows, sleeping monsters. But when material falls in, the release of gravitational energy ignites the infalling gas, causing it to glow intensely as it crushes in toward oblivion. The swirling gas is a mad rush of plasma, with charged particles whipping furiously around from the tremendous forces. Generating loops of self-reinforcing magnetic fields, some material can eject itself to safety before crossing the surface of blackness that marks the outer edge of the black hole itself. The ejected material can stretch for thousands of light-years, powered and sustained by those same twisting magnetic fields, piercing beyond the host galaxy and into the surrounding cluster itself.

Thus, an *active* galaxy, the most powerful engine known in the universe, among the most energetic events since the big bang itself, driven by the extreme gravities of the true monsters of the cosmos: the black holes. When galaxies are active, they scorch themselves with radiation, heating their own gas and preventing the cooling of nebulae, slowing down star formation. Thankfully, the Milky Way is dormant now, as are most galaxies in the present-day cosmos. But in the distant, more crowded past, when larger structures like clusters first started forming, galaxy mergers were much more common, and these active galaxies were much more numerous. It was dangerous times then, and much louder in the radio—these were the bright but distant sources seen in the 1950s that started putting the screws to Hoyle's steady-state model and early credence to the big bang that we all know and love.

Every galaxy is built from the successive mergers of smaller ones, with many repeated feedings of the central black holes (which you'll be unsurprised to learn also grow as more galaxies combine). Eventually, though, for most galaxies, the mergers cease, and they can begin to live in peaceful bliss, resulting in the development of elegant spiral arms. That's right: it's thought that spiral arms are the *natural* state of a galaxy when left to its own devices. Density waves spread through each galaxy like ripples in a pond, but unlike a pond, the galaxy is spinning with its inner bits faster than its outer bits, causing some of the waves to pile up on each other in, well, a spirally fashion. The density differences in an arm are pretty mild—just a few percent—but can trigger the collapse of nebulae and lead to a new round of star formation wherever they occur. Hence the beautiful spiral patterns. It's not that the arms are much more populous than the galactic average, but their demographics are shifted toward recently formed young, blue—and blazing hot—stars. You know, the ones that stand out in visible light photography.

Crash two spiral galaxies together, and it makes things messier. But "crash" is really the wrong word. Merge? Meld? Combine? Galaxies are so much empty space that even the vast nebula clouds are bare flecks of dirt in the galactic landscape. When galaxies collide, it's more like two swarms of bees coming together (and yes, I'm using the same analogy as I did with the larger galaxy clusters, because it's a good analogy). The individual stars won't collide (for the most part; you just know someone's going to be obnoxious, though), but all the gravitational interactions will tear the arms apart, leading to a torn and tattered mix—an irregular galaxy. Over enough time the galaxy heals from its wounds and settles down, but the collision triggered a flash of new star creation, exhausting a galaxy's supply of usable gas. In the long term, the result of a major merger event is a dead (or at the very least, dying) galaxy, full of old red stars: an elliptical.

So galaxy *type* might be connected to galaxy *history*, and this history is driven by the underlying secret machinations of dark matter. Perhaps that's another clue we can use to tease out the dark parts of our universe by performing autopsies on the post-wreck galaxies among us.

Or not. Baryonic processes are enormously complicated, which is good (life) and bad (cosmology), so it may be too difficult to learn anything of bigger value. But no matter, we'll leave that problem for other scientists.[6]

Within and even around galaxies, hydrogen finds ways to clump together; to form stars; to fuse to heavier elements, spread back out, and repeat the story. Repeat this process ad nauseam across the universe, and voilà: a cosmos full of stars, galaxies, light, and vitality. A universe happily churning out generation after generation of stars, each galaxy a factory, inundating the cosmos with light and warmth for the billions of years since the awakening of the cosmic dawn: we live in the stelliferous era, the star-forming and star-loving age of cosmic evolution. An age when elements are fused and the star-nebula cycles continue for generation after generation. An age when life can establish footholds on planetary surfaces and find nourishment for billions of years. An age when the vast array of complicated, messy, but beautiful and energetic processes can play out their parts, each star a voice in the grand cosmic symphony.

An age that is already dying.

CHAPTER 13

THE FALL OF LIGHT

T he next time you see someone on the street corner shouting that the end is near, you can correct him. The end is not near—it is already upon us.

Our Milky Way galaxy currently pops out a litter of around ten stars every single year, with a variety of masses. The majority are small, because small = easy, but there are a couple of medium ones like our sun, and every once in a while a blazing giant. This process has been pretty steady for a couple billion years or so, but it's actually on the downswing. Our home used to be much more efficient at manufacturing stars; nowadays it's relatively lazy compared to its more productive years.

The ability of the universe at large to convert random blobs of gas and dust into stable nuclear reactors peaked long ago—more than nine *billion* years in the past. Before dark energy came into prominence, before structures even finished coalescing, the fat lady was already onstage warming up her vocal cords. It's been downhill ever since, with fewer and fewer stars coming online every day.

Indeed, the majority of stars *ever* to be born in the entire history of the universe, both past and future, have already formed. When it comes to the stellar output of our universe, this is all we're gonna get, folks.

We're both pretty sure and not exactly sure why the lights are going out. We know that the universe is expanding—and accelerating, too—so that's going to slow down the formation of structure, preventing the continual infall of new gas reserves into the galaxies. But star formation is sort of a complicated process, depending on lots of supercomplicated factors, so while it's no surprise that we're over the hill, we don't exactly understand why our downward slope is so steep.[1]

Still, our sun continues to burn, happily converting hydrogen into helium in its infernal core, providing warming light to Earth and all the other planets that don't care as much. Our solar parent is about halfway through its life cycle. There's plenty more hydrogen to spare inside the sun, but most of it won't make its way down to the core where it can do something useful. In about five billion years, the sun will bid farewell to the main sequence and begin its transformation into an ugly red monster.

But the Earth will be doomed long before then. The sun was dimmer in the past, to the dinosaurian delight. The uncomfortable consequence of that little nugget is that the sun will be hotter in the future. Imperceptibly it grows brighter. Not over the course of centuries or decades (sorry, this isn't the source of climate change), but over much longer timescales. In a few hundred million years, give or take, the sun will brighten to the point that the Earth's atmospheric temperature will trigger a runaway greenhouse effect, boiling the oceans.

Unrelatedly, as Edmund Halley didn't realize he realized, the moon is slowly spiraling away from the Earth. In the same amount of time, it will be too small in the sky for total eclipses to occur. So that's a handy astronomical bellwether: when eclipses stop, it's time to get packing. You may also have noticed that the oceans are boiling, but still.

While *life* on Earth will take a hit—microbes clinging to meager hope in polar ponds and subsurface streams—the body of Earth will be just fine, just a tad warmer, joining the rest of the planets in joyless rockishness. The so-called habitable zone, where Goldilocks finds it *just right* for liquid water to exist on a planetary surface, will steadily move outward, warming Mars and, in the very far future, the ice-locked moons of the outer worlds, providing a new, but relatively short-lived, home for life. If humanity is (a) still around and (b) emotionally/physically attached to the Earth, we could soothe the burning heat by lassoing an asteroid, sending it in a looping orbit to gently tug the Earth outward through the millennia and keep us right in the sweet spot. But that is definitely Somebody Else's Problem.

Over the course of a few billion years the Earth's core will cool down and turn solid, switching off our planetary defense screen (a.k.a. the magnetic field), rendering our atmosphere exposed to the onslaught of the solar wind. Plate tectonics will shut off too as the interior cools, eventually making the Earth, once vibrant with life, a wasteland.

Due to tiny gravitational interactions with Jupiter, Mercury's orbit is unstable and chaotic on long timescales. Within a few billion years it has a decent chance of simply ejecting from the solar system with no more warning than a "Later, guys!"[2]

Assuming we don't hitch a tow cable to the Earth and it stays in its current orbit, the onset of the sun's red giant phase is the offset of the Earth. When the sun swells it *swells*, with the outer tendrils of the engorged atmosphere reaching to our orbit. Once embedded in the outer layers of the demon sun, our home planet (and perhaps the only planet we lonely humans will ever know) will have a measly fifty days before friction drags it inward to complete, crushing oblivion. The only sign that our entire planet even existed—with its abundance of life and history and civilization and parking lots—will be a minuscule elevation of the metal enrichment in the star that killed its most beloved child.

There's a small chance that the temperamental fluctuations, mass loss events, and other outbursts during this twilight phase of the sun's life will simply destabilize, rather than consume, the Earth, allowing it to persist as an overroasted marshmallow orbiting the increasingly decrepit sun. So there's that small comfort if you need to sleep tonight.

While the nuclear drama unfolds in our solar system, a greater performance of two massive lovers plays out on a much larger stage. At about the same time our sun breathes its last, the Milky Way and Andromeda, racing together through the vastness of empty space though all these eons, will finally—and tragically—embrace. At first it will just be a touch as the outer spiral limbs encounter each other. But the headlong rush, as paramours across the world know, is inevitable. Over the course of hundreds of millions of years, just as our sun settles into long-term white dwarf retirement, the galaxies will merge into one. There will be multiple passes as the remnant galaxies swing back and forth through each other before finally settling down, and each encounter will trigger a new round of fresh star formation as gravitational ripples send shock waves racing through the once-quiescent nebulae.

But that embrace comes at a price. The onrush of new stars eats up the available supply of gas in a flash. Isolated, the galaxies could have steadily produced new stars far into the future without exhaustion, but fueled by hormones and newfound feelings, they drive their star formation to rates not seen since the reckless early days of cosmic history. The combined galaxy, finally merged after these countless years of eager anticipation, eats itself alive from the inside out.

If our cosmological near-term fate is a bit maudlin, well, I hope you're sitting down, because from here it only gets worse.

But first, a word of caution. The events I've foretold are coming relatively soon and are relatively certain to happen. We see what happens when our sun's cousins stop fusing hydrogen, because there's been enough time in the universe for us to be able to see it with our telescopes. We know what happens when galaxies collide, because the universe is littered with those galactic wrecks, and we've confirmed observationally that directions are locked in—Andromeda and the Milky Way are doomed to that fate.

These events will all play out over the next few billion years—not much longer than the baker's dozen that the universe has already been around. From the perspective of the intense exotic and nuclear ages, the current universe is impossibly old and cold. From the perspective of what I'm about to relate to you, our current epoch is barely a cosmic toddler. Just as how when we dug into our ancient past, the physics became increasingly murkier, so too will our predictions of the ancient *future*.

Our universe was outright bizarre billions of years ago. As we trundle on, it will become equally bizarre, inhabited by creatures we can only dream of now. Deep time plays deep tricks, and as the great cosmic web unravels itself from its current splendor, its constituents—the dark matter, the galaxies, the stars—will change and evolve. Over the course of *hundreds* of billions of years, far longer than the universe has currently existed, our cosmos will become almost entirely unrecognizable.

The story starting a hundred billion years from now is based on our current understanding of the universe, as flawed and possibly myopic as that may be. The further we go into the future, the weaker our confidence gets. To avoid littering the account with *maybe* and *we think so* and *this one cosmologist estimates based on some questionable calculations*, we need to make a deal. I'll start with the future history of our universe, assuming that dark energy continues to do its thing (accelerate the expansion of the universe), dark matter doesn't act up and do something weird, and all the physics we know hold across space and time. After that, we'll talk about some interesting alternatives.

But it doesn't really matter: every possible fate of the universe we can concoct, constrained by known observations, is equally miserable, just in its own unique, quirky way. It's the ultimate price for living in a changing, evolving universe. Once we realized that our cosmos is not static, that it was different in the past, we would inevitably begin forecasting its future. What an ironic twist from the dreams of Kepler: instead of the stars determining our fortunes, it's us who tell the stars their fate.

So let the misery commence.

Dark energy is cruel and implacable: it drives the expansion of the universe to ever-faster speeds without remorse or regret. Galaxies at the distant edges of the universe are already inaccessible to us—they are receding faster than the speed of light, which means even with the most powerful rocket conceivable, we could not hope to move there. The image of that galaxy is formed from light launched long ago; the actual object is now far removed from our reaching grasp. This is a normal and unsurprising feature of an expanding universe, but dark energy makes the whole thing worse, eventually causing even galaxies that are nearby (cosmologically speaking) to be wrested from our view.

Within a hundred billion years—ten times the current age of the cosmos— the entire observable universe will become . . . unobservable. Like losing your eyesight in old age, our view of the celestial realm grows further, redder, and dimmer. The distant galaxies that now populate with glee our deep-sky surveys won't *technically* disappear, but they'll grow so faint and so far redshifted that no telescope could ever distinguish their fading light from the background.

Virgo. Norma. The Great Attractor. Laniakea. The familiar structures of our nearby universe that have been slowly coalescing for billions of years will grind to a halt and reverse, the unmitigated outward pressure of dark energy driving them in the opposite direction. Only the truly gravitationally bound structures will remain. Our own Local Group, dominated by the triumvirate of the Milky Way, Andromeda, and Triangulum, together with a retinue of smaller dwarf galaxies, will persist, holding fast against dark energy's persuasive suggestions. Other clusters and groups will form their own island universes, separated from each other not just by vast gulfs of vacuum but by the limitations of the speed of light.

Within a few hundred billion years, the remaining members of the Local Group will complete their merger, initiated in the now dim and distant past, forming a single large but deformed galaxy, alone in the night.

Within that same time, the cosmic microwave background, that rock-solid bastion of evidence for the big bang, will fade into literal obscurity. It's already pretty cold—just a few degrees off absolute zero—but, lucky for us, clearly visible in the microwave. But with a couple hundred billion more years under its belt, the CMB will be exhausted, redshifting to frequencies so long that not even a telescope the width of the observable universe could detect them. It will, for all intents and purposes, vanish.

If any new life arises in those distant and dark times, they perhaps will never know of their true heritage. No relic radiation to signal the big bang. No distant galaxies to measure a cosmic expansion. They'll think they live in an essentially static universe, unchanging in space and time. If any Fred Hoyle analog arises in that civilization, he'll be right. The cosmological models of our past, of a relatively small universe encompassing the entirety of the galaxy, fixed in time, will turn out to be observationally indistinguishable from the expansion that we know today.

They might get lucky if they're patient observers, perhaps by noticing a peculiar pattern in the redshifting of stars ejected from their galaxy, and deduce that their cosmos was at one time far smaller. And the populations of stars remaining in that single, lone galaxy will indicate that perhaps the universe was different in the unimaginable past. But the universe as we have come to know it will be as alien and unfamiliar as the Planck epoch is to us now.

It makes you wonder if we're missing anything big with our current observational limits, but that's a little too uncomfortable to ponder, so we'll leave it at that.

Indeed, by this time stars as we know them may become a thing of the fanciful past: a tale of a distant golden age to thrill and excite the youngsters around the campfire. As we make the big leap from measuring time in billions of years to trillions of years, the remaining megagalaxy that sits alone in its otherwise empty universe begins to change from the vibrant, rich, colorful panoply that we know today.

For one, stars like our sun will slowly stop appearing and altogether die out. A life span of ten billion years may seem like a long time—and, don't get me wrong, it is—and for us in the present, that's a good fraction of the age of the entire universe. But ten billion is just 1 measly percent of a trillion. As the cosmological eons stretch out to incredible lengths, sunlike stars are as effervescent as fireflies on a warm summer night. The most massive stars, capable of detonating in brilliant supernova explosions, could be missed in a blink of the cosmic eye. And they'll become more rare overall—as star formation slowly grinds to a halt, the capacity for a galaxy to manufacture such massive nuclear beasts dwindles. There simply isn't enough gas in the right places at the right times to build them.

It will be the death of color. White and blue stars will, one by one, disappear from the cosmic stage. Abundant red and blue nebulae will disperse. With nothing left to spin them, the delicate filigrees of planetary nebula will dissolve. The only remaining stars will be small, cool, and red. Indeed, for every star you see in the night sky, there are a thousand in the same patch that you can't. Your eyes are liars—they're not telling you about the true nature of the galaxy. Most stars are too wimpy to be seen with the naked eye, but they're ridiculously abundant.

They already dominate the galaxy, and in a few trillion years, they'll be all that's left. Sole inheritors of a denuded empire.

The small stars win out over long times because they're the economy cars of the galaxy. Their low mass means that gravity is less insane than in bigger stars, so fusion reactions occur at a slightly less eager pace. Their interiors are also constantly churning, drawing fresh reserves of hydrogen into the depths, where it can keep the nuclear fire stoked. Proxima Centauri, another star you can't see with the unaided eye despite the fact that it's so *proxima* that it's our nearest neighbor, will keep chugging along for another four trillion years. In the game of galactic races, always bet on the tortoises.

All stars grow steadily brighter as they age, and since these dwarfs are so numerous, the galaxy of the far future will be roughly as bright as it is today, with its output dominated not by a handful of searchlights but by innumerable cheap LED flashlights.

Stranger things begin to appear. As heavier elements continue to pollute the intergalactic waterways, it changes the makeup of new stars. Not only will big stars become more rare, smaller stars—far smaller than the dwarfs of today—will become possible. The more heavy elements a nebula carries, the more efficiently it can cool itself, allowing chunks of itself to squeeze down tightly and make smaller stars—less than a twentieth of the size of our sun, or about half the mass of the current smallest known star.

A star that small is hard to comprehend, because it's simply infeasible in our current epoch. But given enough time, it'll be common. You'd be surprised to find anything else. And with the heavy elements mixing around its bulk, a star that small could potentially reach surprisingly cool temperatures.

I want you to imagine a star with a nuclear fire raging in its heart, but with water ice clouds circling its frozen surface.[3]

Even those stars will eventually sputter out. It's difficult to tell when the long autumn will come to our universe, because as you may have noticed, the formation, lives, and deaths of stars are a little bit complicated.

The most pessimistic scenario gives only a trillion years until the last star in our universe is born. That's essentially no time at all. More optimistic predictions, trying every trick in the book to keep the fire lit, give a scale a hundred times longer. Either way, eventually the great nebulae of our galaxy will be too thin. Interactions that might trigger a rapid collapse and the birth of a new star will be too rare. And when they do happen by random chance, the energetics will be too low to trigger high enough densities for continued nuclear reactions.

It's in that same time frame, a hundred trillion of these cold, dim years, that we expect the last star to finally burn out. The far-future descendants of today's generation may be far smaller, with far feebler nuclear reactions, but the new molecular mixture of later generations of stars can shorten, rather than lengthen, their lifetimes. Still, longevity is meaningless on these timescales. Even a star that can make it to a ripe old age of ten trillion can't compete against the inexorable march of time.

The universe will simply stop caring about stars.

When the last star sputters into oblivion, it will be the last nuclear fusion reaction that the cosmos will ever produce, save for a few increasingly rare, catastrophic collisions between the dead cinders that remain.

Once the stars fall, the universe will be ruled by the *degenerates*.

These are the remnants, the has-beens, the never-weres: the sad, sorry states that befall all stars. When a star "dies," it doesn't necessary go *poof* and vanish (unless, of course, it blows up, which can happen). There's almost always an object remaining. Much smaller and more pitiful than its progenitor, but still there.

A star like our sun will eventually leave behind a carbon and oxygen ball about the size of a planet—a white dwarf. At the present epoch, these are brilliantly hot objects, which makes sense since they used to be the hearts of stars. When they are first exposed, they blast their environs with hard X-rays, but that fades after a mere ten thousand years. They still remain blazing hot for eons, but now we're in timescales where eons come and go with ease. Eventually they cool and solidify, and when carbon turns solid, it naturally arranges itself into interesting crystalline patterns, which you may know by a more familiar name: diamonds.

The smaller stars, unable to turn helium into anything heftier, simply sputter out without much fanfare, leaving behind a lump of inert helium: a shrug, muttering to the universe, "Eh, I give up."

The most massive stars will be long gone a hundred trillion years from now, but their leftovers remain scattered around the ruined, disfigured clump of our galaxy. Neutron stars, the more massive cousins of the white dwarfs, are a couple of suns' worth of pure neutrons (hence the name) crammed into a sphere the size of a city. For both these neutron stars and white dwarfs, they're supported against gravitational calamity not by any nuclear fires but by the simple refusal of electrons and neutrons to cram themselves too tightly together, a wonderfully quantum phenomenon known as *degeneracy pressure*.[4]

Despite their cold hearts—or maybe because of it—they will persist through the coming death of light.

And then there are the black holes. First considered a mathematical curiosity—a freak show generated by general relativity but not found in nature—they turned out to be . . . found in nature.[5] When nothing can fight against gravity, not even the resisting pressures of crowded neutrons or electrons, the insatiable urges of gravity drive everything into an infinitely small point—the singularity—encased in the one-way boundary of the event horizon.

There are the supergiant supermassive ones, like our friend Sagittarius A*, whom we met in chapter 8, and there are a far greater number of smaller ones floating around the galaxy, the remains of the most massive stars after they spent their fuel. Their numbers will only increase with time.

Last are the most pathetic degenerates of all, the brown dwarfs: loose collections of hydrogen and helium, too big to be called planets but too puny to ignite fusion reactions and name themselves among the stars. The galaxy, even today, is littered with these dim, abandoned half-wits. No solar system to call—or make—a home, these vagabonds wander the dark reaches of the galaxy, hardly ever interacting with or even encountering another object for millions of years at a time. And in this era ruled by the degenerates, still they travel, aimlessly drifting among the tattered remnants of what was once an energetic, bustling metropolis.

Any remaining planets, having survived the deaths of their parent stars, have long since been ripped from their decaying homes and forced into odd, random trajectories. While each solar system is extremely isolated—even in our tidy galactic suburbs in the present day, our nearest neighbor star is twenty-five trillion miles away—given enough time (and don't worry, the universe has plenty of that to spare), stars will eventually pass by each other. When they do, the gravitational interactions will pluck an unlucky planet or two from their cozy, stable, familiar orbits and send them flying off to face the deep darkness alone.

Within a quadrillion years, when the white dwarf remnant of the sun has faded to a handful of degrees of absolute zero, no bound solar systems remain throughout the universe. The galaxy/universe is now split roughly evenly between the brown dwarfs and white dwarfs (now more properly called black dwarfs), with a small fraction of rogue planets and a few neutron stars and black holes. That is the entirety of macroscopic objects, all scattered randomly, all completely, totally isolated.

An occasional flash of light illuminates the decay when two degenerates collide, igniting in a brief but intense supernova or flaring star, a reminder of what the universe was once capable of. But this is no more than a handful of embers, a mere echo of the hundreds of millions of burning torches that once enlightened the Milky Way.

Over time the galaxy, alone in its pocket of the ever-expanding observable universe, begins to dissolve. One by one, the same rare near misses that detach planets from their parents occasionally give a stellar remnant a burst of energy, sending it flying away from the galaxy altogether into the vast emptiness beyond. Once a remnant is free from the chains of gravity that kept it orbiting within the galaxy, dark energy can apply its wicked influence on it. And by now, with the universe in such an advanced age, dark energy is by far the greatest force in the cosmos. One little taste is all it takes for the remnant to be ripped away from its home, flung to impossible distances, literally never to be seen again.

Over the course of a few tens of quintillions of years (that's about eight billion times the current age of the universe—our numbers here are quickly growing preposterously large), up to 90 percent of the galactic members will be thus ripped into cosmic seclusion. The remainders, the most massive and the most lucky, that managed to cling to the home that gave birth to them suffer an even worse fate. Their orbits around the central supermassive black hole— now swelled far larger than its present-day size—emit gravitational waves.

Any orbiting body will do so, even here, even now. But gravity, being by far the weakest force of all, doesn't usually enter into our attention except through extremely accurate observations. But once again, given enough time, the universe can make the imperceptible obvious. Orbit by orbit, the remaining objects slowly, agonizingly, spiral in toward the doom. Due to the pathetic strength offered by gravity, this takes an eon of eons, so much so that we've ran out of Greek prefixes.

In a poignant symmetry, we started the story of our universe with exponential notation to express the intense action happening in mere fractions of a fraction of a second. Now, at the opposite end of the life of our universe, processes take so long to complete that we need to return to that notation.

In this case, within 10^{30} years the universe will be composed of only solitary objects. All orbits, whether remnants around the central supermassive black hole or binary pairs, will have decayed. If a brown dwarf or neutron star hadn't managed to escape the galaxy, by now it will have fallen into the gaping maw of the central black hole.

Imagine life arising—or, if you prefer, some form of consciousness clinging to persistence—on one of the surviving rogue planets or brown dwarfs. Since dark energy operates so efficiently in this ancient universe, you are permanently marooned. Even if another object were within the speed-of-light limitations of communication, its light—assuming it even emitted anything—would be so feeble there'd be no hope of detection. Your entire cold, sluggish existence, limited to a few degrees of absolute zero, would be confined to a single, solitary object, completely and crushingly alone in the vast expanse of nothingness that surrounded you. No star, or companion planet, no *anything* to signal the existence of anything else. Unless you had some memory of the distant past, of what used to be, would you even know the rest of the universe existed?

And then things get weird.

CHAPTER 14

THE LONG WINTER

Have you ever lost touch with a longtime friend? Someone so close you were planning on naming your kids after each other, but then a sudden move or change in lifestyle separates you. At first it's not much—you still meet up for drinks. But as the years go on and you focus on other priorities and other relationships, you realize you haven't spoken to her in ages. You don't even keep track of her on social media, man. Before you know it, you're not even sure where she lives anymore, or the name of her kids. Did she even have kids?

It's inevitable, but it's the way of things: once a pair is separated, it's hard to bring them back, especially if dark energy is involved. First the cosmic web dissolves, then the galaxy itself. Each observable patch of universe past 10^{30} years of age consists of a single object, whether it is a brown or black dwarf, a planet, a neutron star, or a black hole. That's *it*. Each macroscopic object with an entire observable universe to call its miserable own.

And over time, even those macroscopic objects dissolve, and it's the physics of the subatomic world that govern their fate.

In yet another symmetry in this story, the exotic, turbulent early eras in our cosmos shaped the eventual growth of the familiar structures in the present-day universe. In the unfathomably distant future, when the universe has groaned into a near-dead, endless winter, those once-magnificent structures unwind themselves, breaking smaller and smaller, eventually returning themselves to the particle constituents that—for a glorious epoch in the history of the universe—once made themselves into something great.

Dust to dust, as they say.

It's not so much that nothing *interesting* happens in the multiple-quadrillion age of the future universe, but that everything that does happen is so much slower and colder and relies more on pure chance than intentional action. All the forces of nature are still there and still operating, but anything unstable (for really serious definitions of the word "unstable") has long since vanished. But now we must adjust our definition of "stable" too. Objects that we consider *permanent* are anything but, given the extremity of the timescales involved. And that's the opportunity for the universe to engage in some, like I said, *interesting* games.

Take, for example, the noble proton. It is by all accounts as stable as your marriage (uh, let's assume). You could hold a single proton in your hand, and you'll get tired of holding it long before the little bugger does anything.

But we're not exactly sure just how stable the proton is over *ludicrously extreme* lengths of time. Remember the GUTs? The wish-we-understood Grand Unified Theories combining the strong, weak, and electromagnetic forces in a single happy home? How at the outset of the big bang, the dissolution of the GUT may have triggered the inflationary epoch? Yeah, good times, good times.

Well, in some of those fancy GUT theories, the proton can just up and vanish if it so pleases. It's just rare enough that we'll basically never see it happen. In fact, it's exactly such processes that might provide an alternate route to explaining why there's more matter than antimatter in the universe, so the vanishing proton is not an altogether crazy notion.[1]

If the proton decays, then somewhere between 10^{36} and 10^{43} years from now (uh, excuse the rather large uncertainty, but the math is starting to get a bit sketchy here), then all macroscopic objects that aren't named "black hole" will simply dissolve. The neutrons aren't long for the universe either, in case you were wondering: the same physics that would transform a proton could also do so to a safely bounded neutron, the unsuspecting fool.

In a bit of a reprieve against the gloom, the gradual decay of protons is able to (gently) heat any old dead stars once more. Not much, just a few hundred watts of heat. So not enough to run, say, a toaster oven, but far, far warmer than anything else that's going on in the universe (i.e., nothing at all). Proton by proton, a white dwarf sheds mass and converts back to its primordial state: pure hydrogen in a relatively dense ball.

It's not *nearly* dense or hot enough for traditional nuclear fusion—that ship sailed a long time ago—but *untraditional* nuclear fusion is still fair game. It's very simple according to the rules of quantum mechanics: put two atomic nuclei together and wait for an exceedingly long time, and by pure random chance they might combine, providing a brief spark of energy in the process.

For reference, through this stage of life, any white dwarfs have a surface temperature of less than a tenth of a degree above absolute zero.

But the grinding continues, eventually reducing the mass of the white dwarf so much that it simply unglues itself—there's not enough stuff to hold it together. A slowly expanding and dissolving thin cloud of hydrogen is its one and only fate, a fate shared by its nuclear star cousins, planets, and brown dwarfs. Not that it would ever find out, since those objects would be in their own personal patches of the universe.

If the proton *doesn't* decay, which is a very reasonable possibility, then the above scenarios still play out by other, more exotic processes. They get to stay as "recognizable objects" for a lot longer, however, which is small comfort, somewhere into the range of 10^{200} years.

I know I'm breezing through all this cosmological history like it's nothing, but I think it's important to remember that these objects, while dim and desiccated, live *extremely, fantastically, overwhelmingly* long lives. With each paragraph, we're leaping from epoch to epoch, jumping multiple multiples of the current age of the universe.

But that's only because we're counting in years, the length of time it takes for the Earth to orbit around the sun. By the degenerate era, there are no Earths orbiting any suns, so this timekeeping device is just another relic of a long-dead age. In general, it makes sense to think of the passage of time as "the interval between interesting events." The cycles of the seasons on Earth are interesting—and regular—enough to qualify, for humans. When we looked at the initial moments of the big bang, interesting things would occur in a tiny blink, but to the frenzied subatomic processes involved, it was several lifetimes.

Now at the far end of the scale in the endless cosmological winter, life in the universe is much slower and much colder than it is today, which makes it *seem* like an eternity between events, the same way that from the point of view of a hypothetical thinker living in the inflationary epoch, our present universe is embarrassingly slothlike. But to the denizens of this far-flung era, life is just . . . normal.

Even the black holes don't make it, given these hilarious lengths of time. This seems like a good time to inform you that as best we can tell, black holes aren't exactly 100 percent black. Just *almost entirely, but not quite* black. Due to a strange and kind-of-understood quantum mechanical process at the event horizon, dubbed Hawking Radiation in honor of the Stephen who figured it out, black holes emit a very tiny amount of light. You'll thank me for sparing you the gory technical details (especially since most popular descriptions of this process don't really get to the heart of the physics),[2] and because black holes are a mere supporting actor in our story.

The prime takeaway is this: black holes emit radiation and lose mass. Slowly. Like, the equivalent of one photon per year slowly. But hey, when you've got 10^{50} years to play around with, you're going to spit out some significant mass. So all the black holes, each isolated in its own special observational bubble of the universe, decay.

Somewhere in the ballpark of 10^{100} to 10^{200} years (but who's really keeping track now?) every single macroscopic object that is or will be formed will be gone. Dissolved, disassociated, disintegrated. Even protons and neutrons, the venerable baryons that were forged in the first dozen minutes of the big bang, will whittle away to other, more fundamental particles.

All that's left, after these countless eons (even though we're trying our hardest to count them), is a cold, thin soup of photons, neutrinos, electrons, and a few stray other fundamental particles. Some positrons inhabit the cold, dark depths, a leftover of the decay of protons. These positrons can "find" (for lack of a better term) a stray electron and bind to it using the force of, get this, gravity. Orbiting a common center of mass at a distance of, say, a light-year, these exotic creatures will be the last higher-order structures known in the universe. Eventually, of course, the orbits of this *positronium* decay, and the particles annihilate each other in a rare flash of light.[3]

Completely unglued from each other, the particles become their own isolated universe, in a repeat of the process that happened for larger objects in the degenerate era. One photon, or one electron, or one neutrino, alone in its entire observable cosmos, slowly losing energy and approaching absolute zero.

With no heat differences, with no hot springs to contrast with cold flows, the ability to do work is nullified. No work means no consumption, no computation, no cognition. If any form of life makes this far, it too eventually grinds to a halt.

The grim endgame: the heat death of the universe.

That . . . that can't be it, right? *That's* the fate of the universe as predicted by modern physics? That's all we get, a slow winding down of energy differences and the dismemberment of structures? We've worked so hard over the past few centuries to plumb the deepest mysteries of the cosmos, and this is all we can show for it?

Sadly, as depressing as this scenario is, it's a simple extrapolation of the physics of the universe as we know it. If you know where your car is and how fast you're driving, you've played the same game to figure out when you'll get to that party. Except there's no party here, just a miserable end to an enfeebled cosmos.

It's easy to get caught up in the melodrama of the long-term fate of our universe because the outcome is just so dang *morose*. I'm guilty of it too, though I don't know what else you were expecting—the title of this chapter is "The Long Winter," after all. Still, we shouldn't get too comfortable with this telling. There are a lot of assumptions and conditionals baked into our current forecasts, and it makes it all seem so safe and predictable and boring and simple.

And if there's one thing the universe has taught us these past few centuries, it's that complexity has a way of taking its revenge.

For one thing, dark energy. The accelerated expansion depends on dark energy behaving like a cosmological constant, applying its accelerating pressure with dumb eternal insistence. But we're not sure if it really is constant—at best we hope to measure it to within a few percent accuracy in the coming decades. Even if we were to apply all our methodological might and constrain the properties of dark energy to one part in—let's go crazy here—a thousand, that still won't be enough. Given the protracted timescales involved in discussing the ultimate fate of the universe, tiny variations can add up.

Indeed, over the past few years some small tensions have arisen between measurements of Hubble's constant when using early-universe probes (like the relic cosmic microwave background) and contemporary-universe tools (like supernova). That's quite possibly explained by operator error on the part of astronomers, but it could be a sign that something fishy is going on in the world of dark energy.[4]

If dark energy is constant, it's not necessarily a *good* thing, so don't get too excited. For example, if dark energy is actually increasing with time—a scenario called *phantom dark energy* because that sounds totally awesome—then the expansion of the universe will become overwhelmingly coercive, ripping apart clusters, galaxies, and even solar systems. In short order, the small patches of vacuum inside atoms tear them apart, dissolving structures in a cascade of doom. If that's the case, then we only have about another ten billion years or so before our entire universe rips itself apart at the seams.

While violent, at least it's quick.

This scenario puts a big red line under the phrase "we don't know what dark energy will do in the future." As time goes on and accelerated expansion continues apace, the universe will look more and more like the inflationary epoch of old. And we think *that* rapid expansion transformed the cosmos and flooded it with all the cool subatomic toys and gadgets that we love today. Perhaps an encore performance is in store for the far-future universe, refreshing and reigniting the feeble flickering candle of our fate?

Dunno.

What about dark matter? You know, the stuff that makes up most of the stuff? If it's anything like we suspect it to be, then it does occasionally interact with regular matter and even itself. Over time, this drains energy from the dark matter particles, allowing them to settle into any nearby gravitational wells, like a brown or white dwarf. Continued interactions can keep them warm (for very minimalistic definitions of the word "warm") through the twilight of the degenerate era. It's not much to go on, but when faced with the *ultimate heat death of the universe*, we've got to count our blessings.

Or the universe could just up and change in a flash. Seriously, *poof*, it's gone and replaced with something else. Don't put the book down, I'm not kidding around. Here's why it's a very real possibility: it's already happened! It would just be another phase transition, like the one that sent the strong nuclear force (and before that, gravity) splitting off from the remaining forces. During that energetic, exotic process, the universe transformed from one state with a certain population of particles and fields to a completely different one.

If you lived through that transition, you wouldn't: the forces and interactions that you depended on would be up and gone, replaced with strange and unfamiliar new species.

Here's the kicker: what if the phase transition of the universe isn't done? What if the current universe, with its four fundamental forces and array of leptons and hadrons, isn't the true ground state of the governing equations? What if the universe got "stuck"?

Imagine skiing downhill, racing to the bottom of the mountain—the ground state—and *watch out for that rock!* and taking a tumble, getting jammed on a slight rise. You can see the rest of the mountain below you, but you're not moving. You're *stable*, but only in a *meta* sense—a swift enough kick (avalanche, abominable snowman, I'm not really familiar enough with winter sports for more examples) would send you tumbling down to the true base of the slope. But if that swift kick doesn't come, you can just chill out and relax; you're not going anywhere.

So *maybe*—emphasizing that word as hard as I can—the universe is in just such a state. Metastable, it can maintain its current arrangement of forces and physical constants for a very long time. Indeed, it's already managed to do so for more than thirteen billion years. So everything looks nice and comfy . . . for now. All it would take to trigger a new *vacuum decay* is a random blip or jiggle in the wrong place at the wrong time. Good thing there's nothing in the vacuum of space-time providing a minimum energy level capable of destabilizing the local patch of reality.

Oh, right. Vacuum energy. Microscopic quantum fluctuations. If you the skier started shaking uncontrollably, a violent jerk might send you continuing on your downward way.

Maybe the universe has already done it. In such a nucleation event, just as in any other phase transition, the cosmos reconfigures itself from a single point in an outwardly expanding Sphere of Doom. Traveling at the speed of light, there's literally no way to see it coming. By the time it overwhelms you, it's already replaced all your electrons, photons, and anything else-ons with . . . with whatever comes next, I guess.

Calculations are rough here, since they depend on physics beyond the standard model. If we stick to what we know so far (it's worth a shot, I guess), the stability of the universe depends on the nature of the Higgs field, since that field was involved the last time the cosmos underwent a phase transition, giving us the clean separation between the weak nuclear and electromagnetic forces. And you thought we were done with exotic subatomic physics.

Now that the Higgs boson has been confirmed to exist, thanks to the tremendous rock-smashing powers of our particle colliders, looking at how the Higgs particle behaves gives us insights into its future. Is it done, stable for all eternity in its ground state? Or does it have more room to fall? Current measurements of the Higgs put us right on the line of metastability, which is of little comfort.[5]

Hey, at least the universe isn't *un*stable (but we knew that already).

Maybe the end isn't an end at all. Quantum mechanics teaches us that reality is ruled by random chance. An electron can just so happen to be on the opposite side of a wall the next time you look at it. Two protons can just so happen to cohabitate the same volume, and voilà, you have a fusion reaction. But the larger and less quantumish an object or system, the less you expect it to behave weirdly. I can lean against a wall all day long without expecting to pass through it spontaneously. I can sit on my couch all day long and not occupy its same volume (hopefully).

Even in the not-quantum world, gamblers still rule the day. For example, there's nothing in the laws of known physics to prevent all the air molecules in the room you're sitting in to spontaneously end up crammed into a tiny corner, leaving you to asphyxiate helplessly in the vacuum. The only reason thoughts like this don't keep physicists up at night is that these conditions are exceedingly, exceedingly, exceedingly rare. There are so many more ways for the air molecules to be jumbled around the room compared to the number of ways they can be crammed in the corner that the random jostling and jiggling at the molecular level almost always leads to an air-filled room.

That was the briefest summary of the concept of entropy that I could concoct, so consider yourself spared a more long-winded metaphor. Entropy itself is a way to count the number of ways a bunch of particles can rearrange themselves, and the second law of thermodynamics—that entropy always goes up in closed systems—comes from the fact that there are way more disorderly states (like air spread evenly throughout a room) than orderly ones (crammed into a corner).

When you throw out a possibility like bodies spontaneously jumping through walls or air molecules conspiring against you, a proper physicist would immediately scoff and say, "Pshaw, yes, it's *technically* possible, but not likely in a bajillion years."

Well, now we're dealing with a bajillion years. The absurd and unlikely are bound to happen. Given an infinity of time, anything that *could* happen *must* happen. What does this mean for the long-term fate of the universe? It's hard to say because we're operating far outside the normal bounds of known and generally accepted physics. It could mean that a new universe—big bang and all—simply pops into existence through a new inflationary event triggered by a random fluctuation. That universe would be effectively cut off from its parent, with its citizens blissfully ignorant of what came before their own bang.

This new universe—which might or might not have its own set of physical laws—would eventually lead to the formation of new cosmo-babies, on and on and on. That would imply that *our* big bang was neither the first nor the last of those dramatic events, but simply one bead along an infinite glittering strand of . . . beads, I guess.

It could mean that a random patch of the universe might spontaneously decrease in entropy, so much so that a complex structure—say, for example, something like a brain capable of something like conscious thought—would get to contemplate its lonely existence before subliming back into the mean. Preposterous? Yes, but in ten to the hundred to the hundred years, the preposterous becomes plausible.

If the universe is infinite in size, or at least capable of infinitely generating new inflation events, *and* if matter can only arrange itself in a finite number of ways (which just might be true due to the quantum limits on measurements), then that means that all possible combinations and permutations of arranging matter in the universe have been realized. *Infinity* is a tough concept to deal with, and scenarios like this certainly aren't helping. In this picture, not only is every possible organization of galaxies, stars, planets, rocks, and molecules brought to fruition somewhere (or somewhen), each possibility is realized an *infinite number of times*.

That means there's a literal copy of this exact situation, either of me sitting in my pajamas typing this sentence, or you wearing who-knows-what reading it. If the universe is infinite in size, the nearest copy is somewhere out there, well beyond our observable horizon (thankfully). If the universe is infinite in time, then this scenario has already occurred and is fated to happen again. An infinite number of times. Cripes, this is getting embarrassing.[6]

I'll be the first to admit that this picture is a bit hard to swallow, but we should remember that, well, the universe doesn't care what we think about the issue, and it's the (extreme) logical conclusion if we're to take our most modern theories at face value.

Or maybe we're just wrong about all of it. It's not like it hasn't happened before.

We don't know if inflation is correct. We don't know how the rules of quantum mechanics can be extended to incomprehensible timescales. We don't know if the technology of entropy can be applied to the whole entire universe, let alone over the course of an exceedingly exponential number of years.

And don't even get me started on braneworld cosmologies or string theories or whatever the kids are calling it these days. The more hypothetical the physics, the more room for creative explorations of the end state (states?) of the universe.

Our knowledge of the universe at 10^{100} years isn't much different from our knowledge at 10^{-100} seconds: woefully incomplete. In both cases it's the energies involved. In the young cosmos, the temperatures are so high and pressures so extreme that the physics of the familiar are melded together into some strange chimera that eludes understanding. In the remote future, temperatures are so low and processes so agonizingly slow that the statistical rules that govern our daily lives lose their identity. In both cases the universe is extreme, exotic, and potentially unknowable. At its core, after centuries of searching, we don't know how the universe began or how it will end—or if those are even reasonable scientific questions to ponder.

But at least there is symmetry.

A GAME OF CHANCE

S o here's the score. Our entire lives, our entire existence save a few all-too-brief excursions, are confined to a thin, fragile shell on the surface of the Earth. Space, and all the threatening emptiness and vaguely malevolent vastness that goes with it, is a mere sixty miles away. That's right: sixty miles. One hundred kilometers. By International Agreement of People Who Know These Things,[1] space is just a leisurely hour's drive away, if your car could drive straight up.

The most generous definition of the entire biosphere—the oceans and land, the otters and terns, the people and bacteria, the dung beetles and Douglas firs, the lot of it—puts our livable home at around 1 percent of the radius of the Earth. That's roughly the thinness of the shell of the egg you cracked open for your omelet this morning.

Our home planet is but one of eight (or eight thousand, depending on your definition) planets, the largest of the inner rocky worlds but dwarfed by the outer gas giants. The sun, that great luminous ball of fusing hydrogen, is but one of hundreds of billions swimming through the Milky Way galaxy, of roughly middle size and middle age—nothing remarkable there. It sits near the edge of what's called the Local Bubble, the blown-out cavity of a supernova that detonated long ago. Lying about halfway out from the dense galactic core at a radius of twenty-five thousand light-years, the sun is perched on a small spur splintering off the much larger, but comparatively minor, Orion-Cygnus spiral arm.

The Milky Way too is just one among a vast number of galaxies in the observable universe, numbering between five hundred billion and two trillion, subject to how quality you think the estimates for counting dim galaxies are. It's one sparkling but relatively small jewel embroidering the great cosmic web. A member of the Local Group, a faction with the Virgo Supercluster, which itself is nested within the hierarchy of our universe, just a branch of the grander Laniakea Supercluster.

Our place in the universe. 'Nuff said.
(This and the next seven images courtesy of Wikimedia Creative Commons;
author: Andrew Z. Colvin; licensed under CC BY-SA 3.0.)

The observable universe itself is roughly ninety billion light-years across, with the cosmic web stretching across is breadth and depth. The Milky Way is twenty-five thousand times wider than the distance from the sun to Proxima Centauri; our patch of the visible universe is a million times wider than that. Of course, the *actual* universe is far larger. Perhaps infinitely so, but at the very least . . . well, numbers are already meaningless here, so let's just go with *significantly* so.

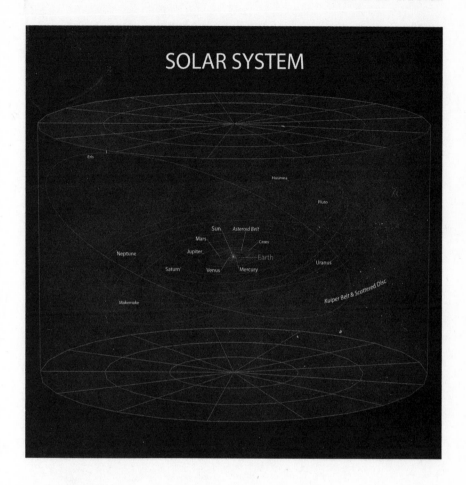

And here we are. After 13.8 billion years of (known) cosmic evolution, from the nuclear maelstrom that birthed the fundamental elements of our existence, the deliberate growth of the galaxy, past generation after generation of stellar births and deaths, comes one particular little star with a family of planets. One of those planets, a blue-colored gem against a backdrop of night, is home to something quite unique and even more surprising in the universe: life.

From the perspective of physical cosmology (which, if you haven't noticed, is the subject of this book), there was no plan, no grand design. The heavens did not single this planet out among all the others. The stars did not whisper to themselves over the eons to conspire and arrange this lucky chance. By all

accounts, we're just here, and the universe had better get used to it, whether it cares about us or not.

But then, we must be a *little* bit special, because where is everybody else? We're still in the early days of needing more than our fingers and toes to count all the planets outside the solar system, but rough estimates land within the ballpark of one trillion for total planets in the Milky Way.[2] That's more than one, on average, per star. Of course most of those obviously aren't good candidates for life (and henceforth I'll use the word "life" to mean "life as we know

it," you know, based on carbon and liquid water and all that. Otherwise we have basically no clue what to look for, so we would have no confident idea of whether we would actually see it even if we had our telescopes pointed right up their . . . never mind, this parenthetical is getting way too long).

Anyway, most planets aren't homes for life. A good number, perhaps most, of those trillion or so planets are unbounded, homeless rogues, not attached to any parent star. Orphaned by ejection events in the chaotic early days of a system's formation, they're doomed to wander aimlessly through the long night. Of the planets lucky enough to call a star home, many are too big, or too small, or too hot, or too cold. The chances of life appearing in any one place are exceedingly, frighteningly slim.

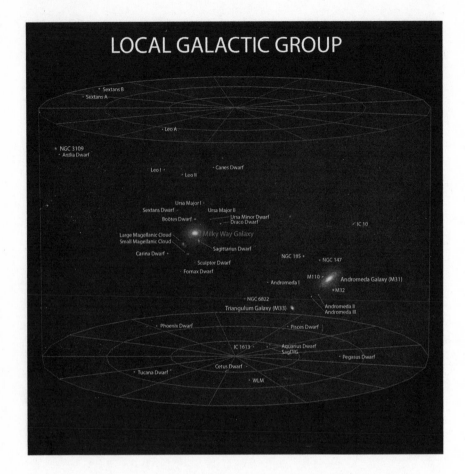

How do we know? Because if life were easy, we would have noticed.

Space is big; space is empty. We've already covered that. But it's also *lonely.* Tens of thousands of detected planets. Probes and rovers and scanners sent to every planet and moon we can reach. Relentless searches for a twin of our Earth circling a distant sun. Countless sleepless nights, monitoring the heavens for the faintest whisper of an alien radio signal.

Nothing. Not a trace, not a hint, not a glimmer. We may not be alone, but we might as well be.

That could change, any day. One day, tomorrow or the next century, we'll catch that whisper, we'll detect that first hint of life, we'll discover a primitive microbe buried under a kilometer of ice. That will truly be a wondrous day,

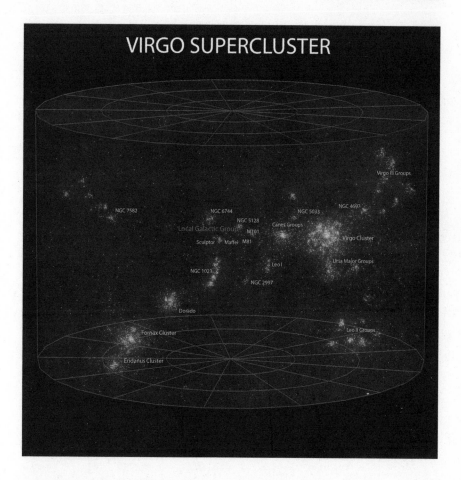

to be remembered throughout the future annals of history: the day we finally proved that there are *others*. That will surely be the first day of a new era for humankind. Or the last. You know, it's a toss-up.

Again, what are the chances for life appearing on another world? I boldly stated that it was slim but not zero, without any, you know, proof to back that up.

Well then, let's rewind.

What does it take for life to appear? What's the right cocktail mix, the right balance of sweet and sour, to get life going? The answer, of course, is "It depends." So far we have access to only one kind of life to study: the life on Earth. Energy from sunlight or deep-sea vents. Carbon for structures. Water

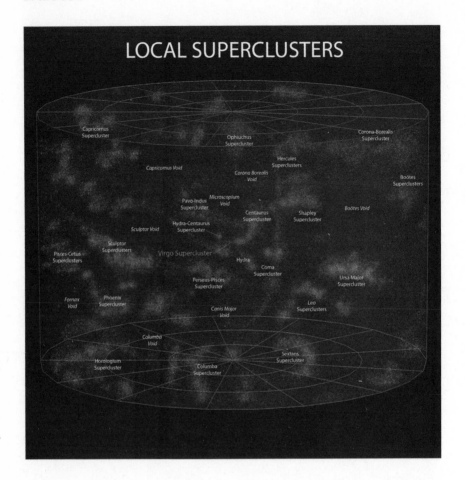

for a solution for chemical reactions. Limited to a narrow range of temperatures and pressures. We may find microbes on the bottom of the ocean and drifting through the upper reaches of the atmosphere, but for all that range, life thrives in just a thin delicate shell. Let's just say that all of the life on Earth could be obliterated and the rest of the universe wouldn't even notice. Do you mourn for that fleck of skin when you scratch your armpit? If you do, you're strange, and the universe is not strange. It wouldn't weep for us.

What did life on Earth need to get the ball rolling? First, it needed liquid oceans. All life on the planet requires water in some way. Life started in the oceans, and the graduation to land happened when organisms could carry their own little bits of ocean with them—the invention of skin.

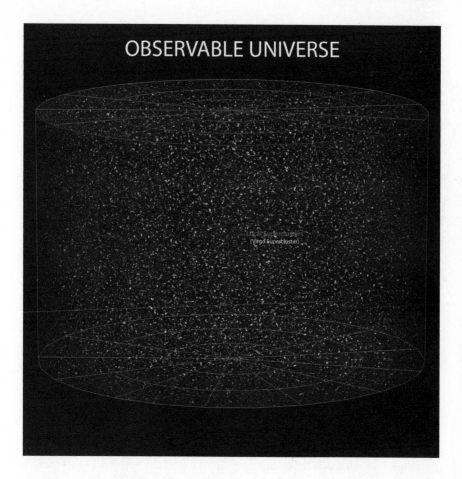

To get a liquid ocean, a planet has to have a heat source. But not too much of a heat source. The sun is a nice, handy heat source. Too far from the sun and all your water's locked up in ice, but too close and the water molecules are too agitated to stick around. Astronomers, always the eager label makers, identify the region around each star where water has the best chances of being, well, water as the habitable zone. Not to be confused with the Twilight Zone or the end zone, it's the ring in each system where Goldilocks finds her favorite soup: not too cold, not too hot.

The habitable zone isn't the only zone, though. Tides can heat up a place too: the constant flexing, stretching, and bending from gravity can warm a world nice and toasty, even if it's far from a sun. To make that work, your candidate planet actually has to be a moon—preferably of a fat gas giant. That's the only way to get enough gravitational tug to turn your core into Play-Doh. Hence all the interest in the (subsurface) liquid water oceans of Europa, Enceladus, and more. But until we know for sure that life has found a holdfast there, we'll leave it to the side here.

Sunlight itself seems kind of useful. After all, it is a giant, constant source of energy available for free to anybody. Some kinds of life on Earth get their energy from other places, like deep-sea hotspots, and while it's currently up for debate whether life got started in those extreme conditions or not,[3] the light-eating kind certainly proved more popular on this planet.

Next, life needs an atmosphere to keep out the constant storm of cosmic rays. I haven't talked about cosmic rays, but for the purposes of this discussion, just understand that the universe is swimming in a constant bath of death-dealing high-energy particles and radiation,[4] and atmospheres make for a great security blanket.

The pressure from the atmosphere helps to keep the water on the Earth . . . on the Earth. Without that pressure, oceans would simply boil away into the vacuum.

Life needs planets with a thick atmosphere, yes, but not too thick! Then the pressures would be too great for delicate life to form extended structures, and the energy of the sun would go into powering storms and winds instead of photosynthesis. Temperatures in those atmospheric pressure cookers can be so high that they can vaporize any water that might have been brave enough to reach the surface. Just look at poor, poor Venus, forever choking in its own haze, too hot and too intense for life to ever get a running start.

Don't forget the magnetic fields. Those cosmic rays are made of charged particles, and magnetic fields can steer them from hitting the precious surface of a planet, either bouncing them away like bullets off Superman's chest or funneling them (relatively) harmlessly into the polar regions. Most planets, especially inner rocky ones, don't get a strong magnetic field. Earth did.

A nice large moon helps, too. See all those craters on Earth's moon? Those are all comets and asteroids that hit the moon instead of hitting the Earth. A few impacts here and there can be a good thing, spurring evolution or delivering some useful compounds. But too much of anything can be a bad thing. One scar on your face can give you character. A hundred? That's a lot of character.

The moon isn't the only thing in the system playing solar goalie for cosmic deathballs headed for the Earth. Jupiter, with its massive gravitational pull, is the sheriff of the outer system, pulling rogue comets into its orbit or kicking them entirely out of town. With a massive planet like Jupiter in the outskirts, the life-bearing inner worlds are that much less vulnerable.

I should note that occasionally Jupiter pitches a rock from the asteroid belt into the inner solar system, so maybe it's a wash there.

Your host star needs to be stable over millions, even billions of years. It can't be young and volatile, throwing energetic tantrums that are no good. It can't be old and senile, filling the inner system with blasts of radiation. Life takes a long time to get its foothold and start running. If the race is over too soon, there's nowhere to go.

And of course, to get water on rocky planets you need (a) some water and (b) a lot of rocks. Not just any old rocks will do: you need good amounts of carbon, oxygen, nitrogen, sulfur, and hydrogen. Life is complex, depending on all sorts of reactions and processes—and those elements are at the heart of them.

You have to pick the right spot in your galaxy too. Too far away from the center, and there aren't enough of those precious elements to make a wet rock. Too close, and you risk being blasted by the intense radiation of the dense stellar neighborhoods.

Taken altogether, the chances of life appearing on any one planet do appear slim.

Really weird orbit? Too bad, no life.

Highly variable star? Too bad, no life.

Too massive giving you a thick atmosphere? Too bad, no life.

Too small for an atmosphere? Too bad, no life.

Not a lot of carbon? Too bad, no life.

No large outer planets? Too bad, no life.

Too much axial tilt and no stable weather patterns? Too bad, no life.

No magnetic field? Too bad, no life.

No plate tectonics? Too bad, no life.

Unlucky massive comet strike? Too bad, no life.

No rotation? Too bad, no life.

Nearby supernova? Too bad, no life.

That's just to get life *started*. The barest, simplest set of criteria needed to add a *bio* in front of *chemistry*. Single-celled organisms ruled Earth for something like a billion years. A billion years! And putting your DNA in a nucleus was once considered a hot new fashion trend. Complex, multicellular life? Land-dwelling life? Life that can bang rocks together? Life that likes to think it can think? That takes *time*. Deep time. Millions upon billions of years of stability. Look how many times life on Earth narrowly escaped complete extinction. How many times the total gene pool was more like a shallow pond.

Given all the opportunities life needs to get started and evolve, it's surprising there's life anywhere at all, let alone intelligent life, let alone life that can write a chapter in a cosmology book about the possibility of life.

And yet, like I said, here we are.

We beat the odds, so far at least. And if we can beat the odds, so can someone—or some*thing*—else.

You most likely did not get in a car accident today. If you did, at least you can read this while you wait to get your arm in a cast. Car accidents are rare, on a personal, individual level. Think of all the tiny little coincidences that have to line up to make *you* get into an accident on *this* drive. Leaving the house fifteen seconds later than normal. Being distracted by that repetitive song on the radio. Looking left at the intersection first instead of right. The sweat on the palm of the other driver's hand, reducing his ability to turn by a few microseconds. The brake pads worn down by 50 percent instead of 49 percent.

Take any one of those elements away, and boom. Well, the opposite of boom. No accident.

The chances of getting in an accident are so low that you don't even think about it. Run out of milk, pop over to the nearest store. Time for practice? Pile up in the back. Date night? Make sure you cleaned out the cheeseburger wrappers.

Despite the chances being so incredibly low, accidents happen *every single day*. Not to you, but to somebody, somewhere. The odds are low but not zero. And that tiny number gets multiplied by the incredible number of cars on the road at any time.

Accidents find a way, and life is an accident waiting to happen.

And so we have a bit of a paradox, named the Fermi paradox in honor of Enrico Fermi, who (naturally) first articulated it.[5] The odds of any planet hosting life are incredibly tiny, almost but not quite zero. So we *do* occupy a privileged position in the universe. Aha, the revenge of Ptolemy! We may not be at the center—that ship sailed a long time ago—but there is something unique, something special about us, about Earth. A little unlike the other planets in the solar system, perhaps the galaxy, and dare I say . . . the universe?

But! Time and time again we've found that we live in a Copernican universe, one where we are not at the center and we are emphatically *not* special. The physics surrounding you right now—the pull of gravity, the photons entering your eyes, the chemistry and thermodynamics—it's all exactly the same across the universe. So life, while rare, can't be *too* rare. If a process or interaction is forbidden in our cosmos, it simply doesn't happen, full stop. If it's allowed, it must be commonplace, because the universe is so freakishly gigantic.

But it looks like life is right on the razor's edge between allowed and not-allowed. Not strictly forbidden in the universe but definitely frowned upon.

So if we're not special after all, and life has a halfway decent shot, where is everybody? Hence, the paradox.

Before resolving it, let's first visit some ways we might be able to spot anybody else.

Freeman Dyson suggested that as we grow up as a species, we would have to go to extraordinary lengths to satisfy our unquenchable thirst for energy. How else are we going to play mind-controlled online poker in the far future? We would find wind power and nuclear power far too wimpy and be inspired to try something truly fantastic: encasing the sun in a giant sphere of rock, collecting 100 percent of that juicy solar output for our nefarious purposes. Of course, such an engine wouldn't be absolutely perfect (even a superadvanced civilization still has to obey thermodynamics), so it would leak a little heat. Actually, a lot of heat. From a great distance, you wouldn't see the star itself (encased in rock, etc.), but you would see something like a blurred-out, surprisingly red, probably infrared, starlike object. If we got overly ambitious, we could do the same to every star we came across, which would be so dramatic it would shift the characteristic hues of the galaxy.

We have found no signs of such constructions, either in the stars we can individually observe or in our deep galaxy surveys. Perhaps that's no surprise when you dig into the details of a so-called Dyson sphere. They require a *lot* of material to build, and you need to spend a lot of energy assembling it—those rocks ain't gonna collect themselves. And as cosmic energy sources go, stars are nice but not that nice. They only last a few billion years (the small red dwarfs are too puny to be worth the effort). Nah, if you were an interstellar civilization on the go, you'd head over to the nearest white dwarf, neutron star, or black hole. Now *those* babies can harness some serious gravitational energy punch.

Or not. We're kind of just making stuff up at this point.

EPILOGUE

Maybe the aliens, whatever they might be, are already here! Look, I have to mention this possibility just so you can't accuse me of not being 100 percent comprehensive, but the chances of alien life actually taking a trip to Earth are so incredibly small under any reasonable understanding of physics that it's so easy to dismiss, I almost forgot to do it. You remember how big-with-a-capital-B the universe is? The incredible distances to even the nearest star? The travel time measured in—at minimum—tens of thousands of years? Distance = time = energy. Colonizing another star, especially with a clunky spaceship big enough to hold some meatbags and their required nutrients, is just about number one on the list titled "Technically Possible but So Infeasible It Might As Well Be Impossible." I'm not one to dismiss romantic thinking, but I'm holding back a serious scoff—and possibly a *pshaw*—at the thought of interstellar travel. It's just not a thing, folks.[6]

OK, so visiting isn't an option. What about just blabbing on the radio? A good old-fashioned chit-chat. We send out a big blast of "Howdy, universe!" on all the frequencies, wait a few dozen (hundred?) years, and get a response back of "How's it going, Earth?" Seems reasonable. Any radio transmitter worth its salt should cut through thousands of light-years of interstellar junk like butter. Our "radio bubble," the ever-expanding sphere of transmissions we've been blasting out ever since we've been able to blast out, is rather small, barely a hundred light-years across. Should someone Out There happen to tune in to the right frequencies, they'll immediately know that something funky is going on, Earthwise.

But any older, or just simply previous, civilizations in the galaxy would have been jammin' for far longer than us, so we should be awash in obviously artificial and obviously foreign radio waves. While we occasionally hear a random bleep or bloop on our radio antennas, they always end up having a rather boring explanation. Reflections from a comet, a new class of unknown star, or even the microwave in the visitor's center (I'm not joking about this one![7]). Even if we couldn't recognize the source of an odd radio signal, aliens are never the answer; a natural explanation, even if it's not completely satisfactory, is always logically preferred over "Aliens did it." Extraordinary claims and so on. You know the deal.

So we've been on both ends of it (the radio blasting and the radio listening) for naught. There are no signs of any superadvanced civilizations reimagining the galaxy with their technological marvels. Nobody's stopped by for a visit. As far as we can tell, and I hope I've made my point clear enough by now, we're alone. What's going on?

Perhaps we're the first sentient species to arrive on the galactic scene. Perhaps there's some sort of filtering action that snuffs out sentient life (whether by self-harm or other, more vague and nefarious, causes).

More likely, we're not comfortable with two things: statistics and large distances.

When it comes to statistics, I suppose I should mention the Drake equation. Drake what? If you're not familiar with it, don't fret. Originated by Frank Drake a few decades ago, probably first on the back of a bar napkin, it purports to quantify the chance of us discovering life, based on variables like the number of stars hosting planets, those planets being in the habitable zone, life surviving long enough to build a radio dish, and so on.[8] The game plan is to make measurements on understanding all the little numbers, and out pops a final probability of getting to make an interstellar handshake. While the Drake equation sees lots of replay action in the discussion on life in the universe, I'm going to be a little blunt here and say that it's absolutely useless.

That's right, I'm going bold: useless.

The Drake equation gives the illusion of knowledge and understanding. You make some assumptions about the requirements for life (like the discussion above, in case you skipped it) and go out making measurements to pin down all those numbers. The problem is that it doesn't really lead to a confident prediction. For example, if you have all the numbers measured to incredible precision and accuracy *except for one*, your final result is still unclear; you have to make precise measures on *all* numbers, or you might as well not have even started. And we have absolutely no way of confidently estimating *most* of the numbers in the Drake equation.

What's more, the very act of *trying* to parameterize ignorance commits you more than you might desire to a particular line of thinking. What if you missed some crucial but nonobvious requirement for life and didn't put it into the Drake stew? You may think you have an answer at the end of the day, but really you're way off the mark. And that doesn't even begin to address the issues of finding life—even life that we might readily recognize—in an unfamiliar and surprising environment. Like, say, the liquid water oceans of the icy moons in our solar system.

In the end, you pour a lot of work into fretting over the Drake equation parameters, only to end up with . . . a guess. You could have just started with the guess and moved on with your life. Or not even bothered playing the game.

We honestly have no clue how rare/unrare life is in the galaxy, let alone the universe. The chances of life appearing on any planet are obviously not zero and also obviously not extremely large. The ultimate answer to why nobody else appears to be home is probably very mundane: life is somewhat common, but intelligent life is rarer, and space is big.

There very well could be at least one other intelligent species hanging out on some rock or two within the Milky Way. But we probably haven't heard from them, or any other past civilizations, because sending radio signals is simply hard. Our own radio bubble, hundreds of light-years across, isn't even distinguishable from the background hum and hiss of the galaxy at the distance of *our nearest neighbor.* In the interstellar regime, a loud and clear shout very quickly just ends up looking like another bit of noise.

Also, the galaxy is *huge.* Gigantic. Supremely large. Other synonyms would be appropriate, but I think you get the idea. And it's constantly evolving, with new stars appearing on the scene and others dying. Perhaps it's just the case that the galaxy is far too large and far too complex for any species to "colonize," even if they really wanted to. Civilizations will appear, grow, decay, and die, making their mark in their little neighborhood and accomplish nothing more.

The Drake equation won't ever give us solid numbers to go on, so we have nothing specific to predict, and of course I can't say anything more than the guesswork offered above. With our searches for planets outside the solar system, with enough sleuthing, we're bound to find a planet with an oxygen-rich atmosphere, a smoking gun that photosynthesis got its game on there, meaning life has found another home. While I'm sure we will celebrate the day that we find life outside the Earth (whatever form it takes), there won't be much to do after that. Back to business as usual.

This line of thinking leads to even more unsettling questions that, if we're going to take our cosmological jobs seriously, we're going to have to confront. It's one thing to talk about the chances of life appearing in our universe at or near the present epoch, with its particular blend of elements and stellar activity. Thoughts along that road lead to some puzzling and partly contradictory answers. But at the next level of existential brainteasers sits something even more critical: why is life even possible in our universe? Like, *at all?*

Look at it this way. Depending on how you arrange them, there are about one or two dozen raw numbers that govern and control all the fundamental physics and cosmology that we know about. The speed of light. The charge of the electron. The strength of gravity. The amount of dark energy. These numbers are like the director of a classic movie. In the finished product, watching the actors emote and dialogue on the screen, we the audience don't get to hear the director shaping and guiding their performance. But take away the influence of that director—or change their attitude or personality—and you get a completely different movie. Sometimes an unwatchably bad one.

Let's say that tomorrow the universe grew tired of having the electron be of a certain charge and decided to double it. Do you think atoms would behave the same way? Molecules? Chemistry? Do you think you would still be alive? Would stars still shine with nuclear fires in their hearts? Would we even recognize the cosmos?

What if there were four spatial dimensions instead of three? Who decided *that*? Would light and gravity propagate in the same way, or would it diminish in intensity so quickly that nothing would ever feel the radiant heat of another object?

What if gravity were stronger or weaker? It wouldn't just affect our ability to get out of bed. Would large structures still form in the cosmos, with reservoirs of gas and dust driven to forge new stars, creating the heavy elements necessary for life?

Dark energy is especially suspicious. We live in a very special time, when dark energy is strong but not too strong, when accelerated cosmic expansion is just beginning to tear the universe apart, but not disastrously so. Currently, regular matter makes up 5 percent of the energy budget of the universe, 25 percent goes to dark matter, and 70 percent is in the form of dark energy. Aren't those numbers suspiciously similar? In the distant past, when everything was crammed together, it was more than 99 percent matter. In the future, as our cosmic butter gets spread too far out, it will be more than 99 percent dark energy.

Why are we in the middle point? That seems too rare and unique. When physical processes compete, especially over the time and energy scales that we're talking about here, it rarely ends up even steven. When one process dominates, it *dominates*. Dark energy's current density value is very, very close to zero but not exactly zero. What's going on? Did something suppress it but then give up? What process brings a competitor to its knees, to the floor even, but doesn't deliver the *coup de grâce*? If you had to pick a random number for the value of dark energy, you'd expect it to be pretty much anything but its current value. In other words, survey says that dark energy should have long ago ripped apart the universe before planets, let alone life, had even formed.

In short, our leaps of understanding of the cosmology over the past few centuries have led us to examine in a new light an old and familiar question: why do we exist, rather than not exist?

We are not special in a cosmic sense, but the universe seems a little too fine-tuned for comfort. Change a fundamental constant or the nature of the some of the big players on the cosmological stage, and life is simply snuffed out. If we're going to take the bold move of shrugging off articles of faith to explain our existence (a surely unintended consequence of the revolutions of Copernicus, Kepler, and Newton), then, well, how are we going to explain our existence?

The short version is that there may be no (scientific) answer and we're just going to have to deal with that on a personal level. Also, that's a deeply philosophical question. I have absolutely no problem with philosophy as a discipline, and I think there are some valuable routes to understanding our world through that lens (and, if we're going to be fair, the entire endeavor of "science" is really just a particular branch of philosophy that highly prizes empirical, in-your-face evidence and lots of math). But this is not a book on philosophy, and I'm certainly no expert at it, so there's not much for me to offer you there.

I will say, however, that when it comes to questions like this, that we have to be very, very careful. Like, *holding a baby chick in your hands* careful.

It's very tempting to shrug our scientific shoulders and say, "Oh, we're here because we're here." If the universe were any different, there would be no life, no consciousness, and no contemplation and examination of matters cosmological. Statements like this are part of a broader category called the *anthropic principle*. Usually these arguments are cast in the mold of eternal inflation or more exotic string theories: if there are a bunch of *possible* universes, all existing and all offering one particular combination of particles, forces, constants, and all the other junk that we call "the physical cosmos," then most of them would necessarily be lifeless—they don't have the right combo. But we see this particular universe with this particular set of physics because it's the right combination that could make us us.

This feels a little bit empty, like eating a bag of potato chips for dinner. It *kind of* explains the problems we have with fine-tuning, but it doesn't really offer any testable predictions or deeper explanations of how the universe ought to work, which is *kind of* the point of the whole scientific endeavor, so it leaves a lot to be desired. I hesitate to even elevate it to the level of *principle* more than, say, *utterance*. But, like I said, this is getting a bit philosophical, and whether you're comfortable with this concept or not is a decision you're going to have to make on your own. No help from me there, kid.

The part that we have to be especially careful about is in counting our probabilities. Let's say you're at the casino playing a game involving dice, laying down the really big bucks. Feeling the excitement—and maybe a little tipsy—you decide to go all in on a single game. The dice are tossed: snake eyes. Bummer. But being mathematically inquisitive, you start to ponder a way to get yourself out of the doldrums. Given that single throw, that one result, what were the chances of getting that bum result?

Well, if the dice were fair, it's pretty easy to calculate. But what if they weren't? What if the game was rigged?

With just a single throw, it's impossible to tell. Testing for riggedness requires a full statistical study with lots of trials and probably a spreadsheet. With only one result to go on, you'll never know if all the outcomes were equally fair or if the casino slipped in some funny dice to tilt the odds in their favor.

We only have access to our one universe, folks. That's it. If the physics that surround us are the result of some random chance, we'll never be able to calculate how "fair" each kind of universe is. It could very well be that our kind of cosmos really is rare. Or maybe it's super common. Just how shifty is the grand cosmological casino? It's definitely not something we can observe, since other universes are *by definition* not a part of our universe and hence not observable. No data = no progress. You can make all the high-powered vocabulary-stretching arguments you want, but without evidence, they're going to be just that: arguments.

Finally, there's a *lot* (and I wish I could make something double-italic to show I really mean it) that we don't know about the universe. The story of the past few hundred years has been one of continually pushing against the sky, laying mysteries on top of answers on top of more mysteries. Questions that puzzled our ancestors now seem laughably quaint and outdated to us, but at the time, they were deep conundrums that challenged our core notions of how reality operated at a fundamental level.

When the tensions grew too thick, like when the old Earth-centered cosmology just wouldn't agree with the wealth of data pouring out of European observatories, or when the debate over the true nature of the spiral "nebulae" spiraled (sorry) out of control, the resolutions came in the form of new physics or new models of the universe—and usually both.

YOUR PLACE IN THE UNIVERSE

We seem to be at a similar crossroads in our modern era. We've just begun to map out the nature of dark matter and are only beginning to pierce the veil that is dark energy. We've explored with our telescopes and our brains the very cusp of the big bang itself, but earlier moments are shrouded in mystery. We know that our physical models of quantum mechanics and general relativity are incompatible with each other, but we don't have a clear path forward (that snake pit is another book).

We've come almost unimaginably far since the days of Kepler and company. Their search for meaning out of the chaos of our world took an unexpected, and unexpectedly fruitful, turn, uncovering a bounty of mysteries—and beauty—within the cosmos that we call our home. The sky that wheels above us every day and every night is only the first layer of a grand and complex structure, almost alive itself in its energetic dances that have lasted for billions of years.

Countless sleepless nights poring over cold data and wrestling with arcane mathematics have teased out a few of nature's jealously guarded secrets. Observation by observation and theoretical insight by theoretical insight, a path forged by generations of scientists ever eager to look upward and inward, we've revealed the full complexity of the universe as it really is to a level that would frighten Kepler and sicken Galileo, while simultaneously discovering deep symmetries and fundamental forces that operate throughout the vastness of space and through cosmic time—a fact that would delight them.

Ultimately, what is our place in the universe? To Kepler's horror, we *are* at the center—of our observable bubble, but that's only a trick of our vantage point. To his, well, equal horror, we're but a tiny, insignificant speck in a cosmos far vaster than he could have possibly imagined. We're simultaneously—and paradoxically—in the middle of nowhere and at the center of it all.

It's so easy to feel disconnected and separate from this universe of ours, but there is something deeper going on. Just as Kepler assumed for all the wrong reasons, we *are* connected to the cosmos. But the stars don't govern our births. Instead, the physics that rule our lives down here on Earth are the same throughout the vastness of the universe.

A hydrogen atom in the laboratory behaves exactly the same as one on the opposite side of the Milky Way. The same force that pulls an apple from a tree shapes and sculpts the largest of structures. The blood in our veins runs with the ash of long-lost generations of stars.

We've come so far in the past few hundred years. What more mysteries await us? What is the nature of the dark part of our cosmos? What is the true mechanism of inflation? How will our universe evolve, and possibly end? As always, scientists are building new engines to enhance our senses. Giant telescopes, gravitational wave observatories, atom smashers, neutrino detectors buried in the ice sheets of Antarctica, satellites operating at all wavelengths of the electromagnetic spectrum, and the stalwart chalkboards, all at the ready. Prepared to wrestle with nature one more time, to fight for one more ounce of understanding, to push our knowledge just one level deeper.

Our assumptions about how the universe works at large scales, using Copernicus and Kepler as a guiding light, have led us well for centuries. We assume that physics is the same throughout space and time. We assume that the universe is, once you look wide enough, homogeneous and isotropic—the same from place to place. And indeed, like all good scientists, we've put our assumptions to the test time and again. Maybe future work will show that those assumptions are wrong, or that our theories are inadequate to the task. I can only hope that our decedents will wistfully say, "They were *so* close, if only they knew . . ." while at the same time pursuing their own, even more profound, questions. Which is fine: the joy of science isn't in the destination but in the path. Curiosity is its own reward.

All the while, deep questions motivate and drive us ever onward. What is our place? Are we special? Does the universe care about us? Well, *we* can think we're special, and *we* can care about each other. And since we—Earth, life, humanity—are a part of the universe anyway, maybe that's enough.

NOTES

CHAPTER 1. SACRED GEOMETRY

1. There's a chance they may have also been totally wasted. S. M. Russell, "Some Astronomical Records from Ancient Chinese Books (Continued)," *Observatory* 18 (1985): 355.

2. A good starting point for reading summaries and translated texts from this era is Anniina Jokinen, "Medieval Cosmology," Luminarium, January 31, 2012, http://www.luminarium.org/encyclopedia/medievalcosmology.htm.

3. For a fun recounting of the spread of Copernicus's viral idea, see Owen Gingerich, *The Book Nobody Read: Chasing the Revolutions of Nicolaus Copernicus* (New York: Walker, 2004).

4. I'm serious. Ann Blair, "Tycho Brahe's Critique of Copernicus and the Copernican System," *Journal of the History of Ideas* 51, no. 3 (1990): 355.

5. Kitty Ferguson, *Tycho and Kepler: The Unlikely Partnership That Forever Changed Our Understanding of the Heavens* (London: Transworld Digital, 2013), Kindle.

6. Ibid.

7. Indeed, Kepler seems like an eager fanboy writing to a reluctant Galileo. Anton Postl, "Correspondence between Kepler and Galileo," *Vistas in Astronomy* 21, no. 4 (1977): 325.

8. Of course I'm paraphrasing, because Kepler goes on and on about this stuff. For example, see a translation of a letter in Edwin Arthur Burtt, *The Metaphysical Foundations of Modern Physical Science* (Garden City, NY: Doubleday, 1954), p. 48.

9. Go ahead and give yourself a blast of a time by reading the whole hog: Johannes Kepler, *Astronomia Nova*, trans. William H. Donahue (Santa Fe, NM: Green Lion, 2015).

10. Gotta love the guy. Johannes Kepler, *Harmonices Mundi*, trans. Charles Glenn Wallis (Chicago: Encyclopædia Britannica, 1952).

11. These are available online today through the library of the University of Kiel in Germany: https://www.ub.uni-kiel.de/digiport/bis1800/Arch3_436.html.

12. A. Athreya and O. Gingerich, "An Analysis of Kepler's Rudolphine Tables and Implications for the Reception of His Physical Astronomy," *Bulletin of the American Astronomical Society* 28 (1996): 1305.

13. Actually lots of books, and naturally, every author has his or her own agenda. For just an entry point into this saga, try Jerome Langford, *Galileo, Science, and the Church* (Ann Arbor: University of Michigan Press, 1992).

14. Galileo Galilei, *Sidereus Nuncius*, trans. Alvert Van Helden (Chicago: University of Chicago Press, 1989).

15. I absolutely need to mention that Brahe had a certain flair for his book titles: hammering the point home, he announced these particular findings in *Concerning the Star, new and never before seen in the life or memory of anyone*.

CHAPTER 2. A BROKEN UNIVERSE

1. Sadly, Maxwell, as much as I'm a fan of his, doesn't get to appear in our story. Sorry, buddy, you'll have to settle for a note: James Clerk Maxwell, "A Dynamical Theory of the Electromagnetic Field," *Philosophical Transactions of the Royal Society of London* 155 (1865): 459.

2. Once again, I could list a few dozen books on the twisting and complicated paths to understanding the earliest moments of the universe. Many of them are, to put it as gently as possible, highly speculative and borderline philosophical. Not that there's anything wrong with philosophers, but physicists usually make for poor ones, and I urge you to keep a large bowl of salt handy when reading anything on this subject. That said, the study of the newborn universe is simultaneously an examination of fundamental physics, which I'm going to explain in a bit.

3. No matter how you pronounce his name, he's a pretty cool dude. Max Planck, "Über das Gesetz der Energieverteilung im Normalspectrum," *Annalen der Physik* 309, no. 3 (1901): 553.

4. Told you so. For a good review of leading (not necessarily viable) solutions, check out Lee Smolin, *Three Roads to Quantum Gravity* (New York: Basic Books, 2017).

5. That's not much more informative, but GeV is a measure of energy: the amount required to accelerate one electron across a potential difference of one volt. In other words, how much you're going to sweat after making an electron do something it doesn't want to do. It seems arbitrary, but when your job is to make electrons do things they don't want to do—like slam together in a particle collider—it starts to make more sense. A weakly thrown baseball has a few trillion electron volts of energy due to its mass; a thin beam of charged subatomic particles with the same energy can and will punch a hole straight through you, so don't even think you can catch it. Trust me; they put up warning signs and everything.

6. Peter Higgs, "Broken Symmetries and the Masses of Gauge Bosons," *Physical Review Letters* 13, no. 16 (1964): 508.

7. For an entertaining deep dive, it doesn't get much better than Richard P. Feynman, Robert B. Leighton, and Matthew Sands, *The Feynman Lectures on Physics*, vol. 2, *The New Millennium Edition: Mainly Electromagnetism and Matter* (New York: Basic Books, 2011), chap. 34.

8. Alan H. Guth, "Inflationary Universe: A Possible Solution to the Horizon and Flatness Problems," *Physical Review* D 23, no. 2 (1981): 347.

CHAPTER 3. TALES FROM A BEWILDERING SKY

1. As recounted in Willian Stukeley, *Memoirs of Sir Isaac Newton's Life*, transcript, 1752, taken from University of Pennsylvania Online Books, http://onlinebooks.library .upenn.edu/webbin/book/lookupid?key=olbp49182 (accessed October 17, 2017).

2. You can get a copy of Newton's great *Philosophiae Naturalis Principia Mathematica* online, but if you want to hold some genius in your hands, then I suggest Isaac Newton, *The Principia: Mathematical Principles of Natural Philosophy* (Austin, TX: Snowball Publishing, 2010).

3. And you know you are. Edmund Halley, "Some Account of the Ancient State of the City of Palmyra, with Short Remarks upon the Inscriptions Found There," *Philosophical Transactions* 19 (1695): 160.

4. And if you've ever seen a map of how an eclipse can be viewed, you can pretty much thank him. Jay M. Pasachoff, "Halley and His Maps of the Total Eclipses of 1715 and 1724," *Journal of Astronomical History and Heritage* 2 (1999): 39.

5. Halley, "Some Account."

6. Naturally it wasn't that simple—he initially thought it was a comet. William Herschel and Dr. Watson, "Account of a Comet, by Mr. Herschel, F. R. S.; Communicated by Dr. Watson, Jun. of Bath, F. R. S," *Philosophical Transactions of the Royal Society of London*, 71 (1781): 492.

7. For a history of its publication and additions, as well as links to some pretty pictures, visit: "Charles Messier's Catalog of Nebulae and Star Clusters," Messier Catalog, last modified August 12, 2011, http://www.messier.seds.org/xtra/history/m-cat.html (accessed November 15, 2017).

8. This tale and more about Galileo's perplexity are recounted in David White-house, *Renaissance Genius: Galileo Galilei and His Legacy to Modern Science* (New York: Sterling, 2009), p. 100.

9. Also of note is that this book contains one of the most powerful explanations for why we do science, in this case applied to the problem of Saturn's rings: "When we have actually seen that great arch swung over the equator of the planet without any visible connection, we cannot bring our minds to rest." James Clerk Maxwell, *On the Stability of the Motion of Saturn's Rings* (Cambridge: Macmillan, 1859), p. 1.

10. I mean, come on, dude, really? Auguste Comte, *The Positive Philosophy of Auguste Comte: Freely Translated and Condensed by Harriet Martineau*, book 2, *Astronomy, C. I: General View* (Cornell, NY: Cornell University Library, 1896).

11. Edward Harrison, *Darkness at Night: A Riddle of the Universe* (Cambridge, MA: Harvard University Press, 1989).

12. Ann Blair, "Tycho Brahe's Critique of Copernicus and the Copernican System," *Journal of the History of Ideas* 51, no. 3 (1990): 355.

CHAPTER 4. THE DEATH OF ANTIMATTER

1. P. A. M. Dirac, "The Quantum Theory of the Electron," *Proceedings of the Royal Society* A 117, no. 778 (1928): 610.

2. Go ahead, take a crack at it. I'll wait. Erwin Schrödinger, "An Undulatory Theory of the Mechanics of Atoms and Molecules," *Physical Review* 28, no 6 (1926): 1049.

3. We get this constraint from detailed observations of the cosmic microwave background, which I haven't introduced yet, but feel free to get the scoop now. Planck Collaboration, "Planck 2015 Results. XIII. Cosmological Parameters," *Astronomy & Astrophysics* 594 (2016): id.A13.

4. Michael S. Turner and David N. Schramm, "The Origin of Baryons in the Universe," *Nature* 279 (1979): 303.

5. Don't say I didn't warn you. A. Karel Velan, "Quantum Chromodynamics, the Strong Nuclear Force" in *The Multi-Universe Cosmos* (Boston: Springer, 1992).

6. I'll leave that to David Griffiths, *Introduction to Elementary Particles* (New York: Wiley, 2008).

7. Edward Kolb and Stephen Wolfram, "Baryon Number Generation in the Early Universe," *Nuclear Physics* B 172 (1980): 224.

8. Which is a shame, because neutrinos don't get a lot of airtime. Try this out if you do want to follow that lead: Ray Jayawardhana, *Neutrino Hunters: The Thrilling Chase for a Ghostly Particle to Unlock the Secrets of the Universe* (New York: Scientific American, 2013).

9. Speaking of random papers, here's the one that really kicked this idea off: Ralph Alpher, Hans Bethe, and George Gamow, "The Origin of Chemical Elements," *Physical Review* 73 (1948): 803.

CHAPTER 5. BEYOND THE HORIZON

1. The light-year is used as a way to easily communicate to the public the large distances to the star he had just confidently measured. Thanks, dude! Fredrich Bessel, "On the Parallax of the Star 61 Cygni," *London and Edinburgh Philosophical Magazine and Journal of Science* 16 (1839): 68.

2. Henrietta S. Leavitt and Edward C. Pickering, "Periods of 25 Variable Stars in the Small Magellanic Cloud," *Harvard College Observatory Circular* 173 (1912): 1.

3. A great resource for the origins of the debate, papers published summarizing the debate itself, and—the juicy bits—reactions by the attendees is "The Shapley—Curtis Debate in 1920," NASA Astronomy Picture of the Day, https://apod.nasa.gov/diamond _jubilee/debate_1920.html (accessed December 2, 2017).

4. Edwin Hubble, "Cepheids in Spiral Nebulae," *Publications of the American Astronomical Society* 5 (1925): 261.

5. Seriously, the dude was a pretty snappy writer. Edwin Hubble, "A Relation between Distance and Radial Velocity among Extra-Galactic Nebulae," *Proceedings of the National Academy of Science* 15, no. 3 (1925): 16.

6. Fritz Zwicky, "On the Red Shift of Spectral Lines through Interstellar Space," *Proceedings of the National Academy of Sciences* 15, no. 10 (1929): 773.

7. If you want to see how long someone can make it, I invite you to read Charles Misner, Kip Thorne, and John Archibald Wheeler, *Gravitation* (New York: W. H. Freeman, 1973), a.k.a. "The Grad Student's Bane."

8. From the man himself: "Space by itself, and time by itself, are doomed to fade away into mere shadows, and only a kind of union of the two will preserve an independent reality." Hermann Minkowski, "Space and Time," in Hendrik A. Lorentz, Albert Einstein, Hermann Minkowski, and Hermann Weyl, *The Principle of Relativity: A Collection of Original Memoirs on the Special and General Theory of Relativity* (New York: Dover, 1952), pp. 75–91.

9. Albert Einstein, "Kosmologische Betrachtungen zur allgemeinen Relativitats-theorie," *Sitzungsberichte der Königlich Preussischen Akademie der Wissenschaften Berlin* (1917): 1: 142.

10. Ibid.

CHAPTER 6. BATHED IN RADIANCE

1. Simon Mitton, *Fred Hoyle: A Life in Science* (Cambridge: Cambridge University Press, 2011), p. 129.

2. For an accessible summary of the problems with tired light, plus links to the research papers, check out Edward L. Wright, "Errors in Tired Light Cosmology," UCLA Division of Astronomy and Astrophysics, April 24, 2008, http://www.astro.ucla .edu/~wright/tiredlit.htm (accessed October 4, 2017).

3. The steady party got started with Hermann Bondi and Thomas Gold, "The Steady-State Theory of the Expanding Universe," *Monthly Notices of the Royal Astronomical Society* 108 (1948): 252.

4. Robert Dicke et al., "Cosmic Black-Body Radiation," *Astrophysical Journal* 142 (1965): 414.

5. Arno Penzias and Robert Wilson, "A Measurement of Excess Antenna Temperature at 4080 Mc/s," *Astrophysical Journal* 142 (1965): 419.

CHAPTER 7. REAPING THE QUANTUM WHIRLWIND

1. Told you we would come back to him. Max Planck, "Über das Gesetz der Energieverteilung im Normalspectrum," *Annalen der Physik* 309 (1901): 553.

2. Albert Einstein, "Über einen die Erzeugung und Verwandlung des Lichtes betreffenden heuristischen Gesichtspunkt," *Annalen der Physik* 17, no. 6 (1905): 132.

3. Like his (in)famous delta function, as introduced in Paul Dirac, *The Principles of Quantum Mechanics*, 4th ed. (Oxford: Clarendon, 1958).

4. The term "quark" itself came from Murray Gell-Mann basically looking around for a weird and cool name for his recently unearthed theoretical construct. The individual monikers came later as physicists just made up stuff from the top of their heads. Murray Gell-Mann, *The Quark and the Jaguar: Adventures in the Simple and the Complex* (New York: Henry Holt, 1995), p. 180.

INTERLUDE: A GUIDE TO LIVING IN
AN EXPANDING UNIVERSE

1. It only took the heroic efforts of hundreds of scientists and engineers, a lot of money, fancy satellite missions, and independent measurements. You know, science stuff. Planck Collaboration, "Planck 2015 Results. XIII. Cosmological Parameters," *Astronomy & Astrophysics* 594 (2016): id.A13.

CHAPTER 8. BEHOLD THE COSMIC DAWN

1. Starting with Tom Kibble, "Topology of Cosmic Domains and Strings," *Journal of Physics A: Mathematical and General* 9, no. 8 (1976): 1387.

2. So many constraints, so little time. Thanks to Planck Collaboration, "Planck 2013 Results. XXV. Searches for Cosmic Strings and Other Topological Defects," *Astronomy & Astrophysics* 571 (2014): id.A25.

3. Seriously, this is one of the weirdest and most fascinating manifestations of

quantum theory: the empty vacuum of space itself influencing motion. Hendrik Casimir, "On the Attraction between Two Perfectly Conducting Plates," *Proceedings of the Royal Netherlands Academy of Arts and Sciences* 51 (1948): 793.

4. Most notably with the COBE (Cosmic Background Explorer) mission. One of its leaders, George Smoot, went on to win a Nobel for his efforts and wrote a nice book recounting the adventures. George Smoot and Keay Davidson, *Wrinkles in Time* (New York: W. Morrow, 1993).

5. Listen, I'm not just plugging the Planck mission because I played a minor role in the data analysis efforts. It seriously is perhaps the most detailed and exacting astronomical measurement ever taken.

6. To get your feet wet, try Brian O'Shea et al., "First Stars III Conference Summary" (Santa Fe, NM: Proceedings of First Stars III, July 2007).

7. It depends on how much you trust your simulation (the answer is almost always "about as much as I can simulate throwing it") and how much we understand the physics of this era. At the "small" end, we expect masses forty times that of the sun. So still pretty big, but, you know, not as big. Hosokawa Takashi et al., "Protostellar Feedback Halts the Growth of the First Stars in the Universe," *Science* 334 (2011): 1250.

8. Welllllll, maybe. But close enough for our purposes. Abraham Loeb and Steven Furlanetto, *The First Galaxies in the Universe* (Princeton, NJ: Princeton University Press 2013), p. 213.

9. There are even curious relationships between black hole mass and properties of their host galaxies, implying symbiotic coevolution. Kayhan Gultenkin et al., "The M-σ and M-L Relations in Galactic Bulges, and Determinations of Their Intrinsic Scatter," *Astrophysical Journal* 698 (2009): 198.

10. Jonathan Gardner et al., "The James Webb Space Telescope," *Space Science Review* 123 (2006): 485.

11. An example of just one such mission is David DeBoer, "Hydrogen Epoch of Reionization Array (HERA)," *Publications of the Astronomical Society of the Pacific* 129, no. 974 (2017): 045001.

CHAPTER 9. OF MATTERS DARK AND COLD

1. Fritz Zwicky, "On the Masses of Nebulae and of Clusters of Nebulae," *Astrophysical Journal* 86 (1937): 217.

2. Vera Rubin and Kent Ford Jr., "Rotation of the Andromeda Nebula from a Spectroscopic Survey of Emission Regions," *Astrophysical Journal* 159 (1970): 379.

3. Albert Einstein, "Erklärung der Perihelbewegung des Merkur aus der allgemeinen Relativitätstheorie," *Königlich Preussische Akademie der Wissenschaften* (1915): 831.

4. The full story is recounted in Tom Standage, *The Neptune File: A Story of Astronomical Rivalry and the Pioneers of Planet Hunting* (London: Walker, 2000).

5. With gedankenexperiment ("thought experiment") the obvious rival.

6. Mordehai Milgrom, "MOND Theory," *Canadian Journal of Physics* 92 (2015): 107.

7. Constantinos Skordis, "Topical Review: The Tensor-Vector-Scalar Theory and Its Cosmology," *Classical and Quantum Gravity* 26 (2009): 143001.

8. Douglas Clowe et al., "A Direct Empirical Proof of the Existence of Dark Matter," *Astrophysical Journal* 648 (2006): L109.

9. It's that good old-fashioned big bang nucleosynthesis: Ralph Alpher, Hans Bethe, and George Gamow, "The Origin of Chemical Elements," *Physical Review* 73 (1948): 803.

10. George Blumenthal et al., "Formation of Galaxies and Large-Scale Structure with Cold Dark Matter," *Nature* 311 (1984): 517.

11. Gerard Jungman et al., "Supersymmetric Dark Matter," *Physics Reports* 267 (1996): 195.

12. David Weinberg et al., "Cold Dark Matter: Controversies on Small Scales," *Proceedings of the National Academy of Sciences* 112 (2015): 12249.

CHAPTER 10. THE COSMIC WEB

1. For a while, astronomers thought that clusters were the biggest thing and were more or less scattered around the universe randomly, much as we used to think stars filled the universe. But deeper surveys revealed the beginnings of what we now call superclusters, and what really kicked things off was the discovery of the great voids— vast regions of no clusters at all. It was that discovery that led astronomers to think that something big was afoot. Stephen Gregory and Laird Thompson, "The Coma/A1367 Supercluster and Its Environs," *Astrophysical Journal* (1978): 784.

2. Margaret Geller and John Huchra, "Mapping the Universe," *Science* 246 (1989): 897.

3. It's easiest to see this in simulations, where we can probe finer structures without having to deal with temperamental telescopes, as in Miguel Aragon-Calvo and Alexander Szalay, "The Hierarchical Structure and Dynamics of Voids," *Monthly Notices of the Royal Astronomical Society* 428 (2013): 3409.

4. And it still continues to be tested today. Here's a random paper on the subject, pulled out of a hat: Rodrigo de Sousa Goncalves et al., "Cosmic Homogeneity: A Spectroscopic and Model-Independent Measurement," *Monthly Notices of the Royal Astronomical Society* 475 (2018). Available online at https://arxiv.org/abs/1710.02496 (accessed July 13, 2018).

5. There is even an entire galaxy survey devoted to measuring this: "BOSS: Dark

Energy and the Geometry of Space," SDSS III, 2013, http://www.sdss3.org/surveys/boss.php (accessed December 12, 2017).

6. This "bottom-up" way of building the universe is in contrast to a "top-down" style, where giant blobs of gas fragment into ever-smaller lumps that we end up calling galaxies.

7. Brent Tully et al., "The Laniakea Supercluster of Galaxies," *Nature* 513 (2014): 71.

CHAPTER 11. THE RISE OF DARK ENERGY

1. The machinery you need to use a particular set of solutions to general relativity first derived from the mustachioed Alexander Friedmann, "Über die Krümmung des Raumes," *Zeitschrift für Physik* 10 (1922): 377.

2. And for the three-peat, Planck Collaboration, "Planck 2015 Results. XIII. Cosmological Parameters," *Astronomy & Astrophysics* 594 (2016): id.A13.

3. And that number hasn't budged much in the decades we've been measuring it. For example, here's another random paper measuring it: Rachel Mandelbaum et al., "Cosmological Parameter Constraints from Galaxy-Galaxy Lensing and Galaxy Clustering with the SDSS DR7," *Monthly Notices of the Royal Astronomical Society* 432 (2013): 1544.

4. Walter Baade and Fritz Zwicky, "On Super-Novae," *Proceedings of the National Academy of Sciences* 20 (1934): 254.

5. OK, maybe a lot of finagling. The methods are far from perfect and introduce their own source of uncertainty, as evidenced when, for example, it was applied to a mere seven supernova and produced a very inaccurate result. Saul Perlmutter et al., "Measurements of the Cosmological Parameters Ω and Λ from the First Seven Supernovae at z > = 0.35," *Astrophysical Journal* 483 (1997): 565.

6. I present you the two towers of dark energy: Adam Riess et al., "Observational Evidence from Supernovae for an Accelerating Universe and a Cosmological Constant," *Astrophysical Journal* 116 (1998): 1009; Saul Perlmutter et al., "Measurements of Ω and Λ from 42 High-Redshift Supernovae," *Astrophysical Journal* 517 (1999): 565.

7. Dragan Huterer and Daniel Shafer, "Dark Energy Two Decades After: Observables, Probes, Consistency Tests," *Reports on Progress in Physics* 81 (2018): 016901.

8. David Weinberg et al., "Observational Probes of Cosmic Acceleration," *Physics Reports* 530 (2013): 87.

CHAPTER 12. THE STELLIFEROUS ERA

1. David Devorkin, "The Origins of the Hertzsprung-Russell Diagram," *Proceedings of the International Astronomical Union*, no. 80 (1977): 61.

2. Joe D. Burchfield, *Lord Kelvin and the Age of the Earth* (Chicago: University of Chicago Press, 1990), pp. 57–80.

3. Frank Dyson, A. S. Eddington, and C. R. Davidson, "A Determination of the Deflection of Light by the Sun's Gravitational Field, from Observations Made at the Solar Eclipse of May 29, 1919," *Philosophical Transactions of the Royal Society* A220 (1920): 571.

4. Jeanne R. Wilson, "An Experimental Review of Solar Neutrinos," *Prospects in Neutrino Physics Conference Proceedings* (April 16, 2015).

5. Edwin Hubble, "Extra-Galactic Nebulae," *Astrophysical Journal* 64 (1936): 321.

6. We'll leave that for scientists like these folks: Mark Vogelsberger et al., "Properties of Galaxies Reproduced by a Hydrodynamic Simulation," *Nature* 509 (2014): 177.

CHAPTER 13. THE FALL OF LIGHT

1. Piero Madau and Mark Dickinson, "Cosmic Star-Formation History," *Annual Review of Astronomy and Astrophysics* 52 (2014): 415.

2. Jacques Laskar, "Large-Scale Chaos in the Solar System," *Astronomy & Astrophysics* 287 (1994): L9.

3. As you might imagine, there isn't exactly a lot of research on the long-term fate of stars and galaxies, if for no other reason than the simple fact that there aren't going to be any observations—at least for a while—to test any hypotheses. Thus the following reference is the go-to standard for most of this story, and in the decades since its publication, there haven't been any major complaints or corrections, except that the authors didn't know that we live in a universe full of dark energy, which does modify the story. Fred Adams and Gregory Laughlin, "A Dying Universe: The Long-Term Fate and Evolution of Astrophysical Objects," *Reviews of Modern Physics* 69 (1997): 337.

4. This phenomenon was first figured out by the supremely talented Subramanian Chandrasekhar, "The Maximum Mass of Ideal White Dwarfs," *Astrophysical Journal* 75 (1931): 81.

5. Naturally, black holes have a long and storied history worth retelling in another book. Their origins, however, are quite mundane: they appear in one of the simplest solutions of general relativity: Karl Schwarzschild, "Über das Gravitationsfeld eines

Massenpunktes nach der Einsteinschen Theorie," *Sitzungsberichte der Königlich Preussischen Akademie der Wissenschaften* 7 (1916): 189.

CHAPTER 14. THE LONG WINTER

1. There's continuing and ever-evolving research on this topic, but a solid review can be found in Antonio Riotto, "Theories of Baryogenesis," (lecture; Summer School in High Energy Physics and Cosmology, Trieste, Italy, June 29–July 17, 1998 [1999]).

2. Sigh, here we go. The usual story is that a particle-antiparticle pair appears in the vacuum of space near an event horizon, with one on the wrong side of the line. It's consumed by the black hole while its partner runs off scot free. This is a "bonus" particle given to the universe, so the energy has to come from somewhere—hence, the black hole loses mass. While this isn't a technically wrong story, I don't think it really represents the underlying mathematics, which is more about the relationship between quantum fields (remember those?) and the sapping of energy from a forming black hole, which leads to its eventual dissolution down the road. But whatever, don't take my word for it. Just read Hawking's original paper on it: Stephen Hawking, "Black Hole Explosions?," *Nature* 248 (1974): 30.

3. And the award for most clever article title in these notes goes to Don Page and M. Randall McKee, "Eternity Matters," *Nature* 291 (1980): 44.

4. Wendy Freedman, "Correction: Cosmology at a Crossroads," *Nature Astronomy* 1 (2017): id. 0169.

5. Alexander Bednyakov et al., "Stability of the Electroweak Vacuum: Gauge Independence and Advanced Precision," *Physics Review Letters* 115 (2015): 201802.

6. If you want to go down this particular rabbit hole, you're going to have to follow Max Tegmark, "The Multiverse Hierarchy," in *Universe or Multiverse?*, ed. B. Carr (Cambridge: Cambridge University Press, 2007).

EPILOGUE: A GAME OF CHANCE

1. It's the Karman Line, a nice round number close enough to the height where the atmosphere is so thin that normal airplane physics doesn't work so well anymore. Dennis Jenkins, "Schneider Walks the Walk; Extra Feature: A Word about the Definition of Space," NASA, October 21, 2005, https://www.nasa.gov/centers/dryden/news/X-Press/stories/2005/102105_Schneider.html.

2. You know, plus or minus a few hundred billion. Takahiro Sumi et al., "Upper

Bound of Distant Planetary Mass Population Detected by Gravitational Microlensing," *Nature* 473 (2011): 349.

3. Rachel Brazi, "Hydrothermal Vents and the Origins of Life," *Chemistry World*, April 16, 2017, https://www.chemistryworld.com/feature/hydrothermal-vents-and-the-origins-of-life/3007088.article.

4. Dimitra Atri and Adrian Melott, "Cosmic Rays and Terrestrial Life: A Brief Review," *Astroparticle Physics* 53 (2014): 186.

5. Seth Shostak, "Fermi Paradox," SETI Institute, April 19, 2018, https://www.seti.org/seti-institute/project/fermi-paradox (accessed December 8, 2017).

6. Before you jump on me, I should say that of course interstellar travel is possible. Objects travel from system to system in our galaxy all the time, and we humans have even hurled a few chunks of metal out into the interstellar wastelands. But what we usually mean by "travel"—the same way we might travel by train or plane to another city—is so far beyond the energy generation capabilities of our civilization, and projections of said capabilities into the far, far, far future, that we might as well discount it as a feasible process for all intents and purposes. And it may never be feasible, even if we could harness unimaginable amounts of energy. In short: you're not going to another star, and neither are your kids' kids' kids' kids' kids' kids. You can probably safely add a few more generations onto that last sentence. Space is big; don't mess with it.

7. Emily Petroff, "Identifying the Source of Perytons at the Parkes Radio Telescope," *Monthly Notices of the Royal Astronomical Society* 451 (2015): 3933.

8. "The Drake Equation Revisited," *Astrobiology Magazine*, September 29, 2003, https://www.astrobio.net/alien-life/the-drake-equation-revisited-part-i/.

SELECT BIBLIOGRAPHY
AND FURTHER READING

Adams, Fred C., and Greg Laughlin. *The Five Ages of the Universe: Inside the Physics of Eternity*. New York: Free Press, 2000.

Bartusiak, Marcia. *The Day We Found the Universe*. New York: Pantheon, 2009.

Berlinski, David. *Newton's Gift: How Sir Isaac Newton Unlocked the System of the World*. New York: Free Press, 2000.

Carroll, Sean. *The Particle at the End of the Universe: How the Hunt for the Higgs Boson Leads Us to the Edge of a New World*. Boston: Dutton, 2013.

Cox, Brian, and Jeff Forshaw. *The Quantum Universe: Everything That Can Happen Does Happen*. London: Allen Lane, 2011.

Davies, Paul. *The Eerie Silence: Renewing Our Search for Alien Intelligence*. Boston: Mariner, 2010.

Ferguson, Kitty. *Tycho & Kepler: The Unlikely Partnership That Forever Changed Our Understanding of the Heavens*. New York: Walker, 2002.

Feynman, Richard P. *The Character of Physical Law*. Cambridge, MA: MIT Press, 1964.

Garrett, Katherine, and Gintaras Dūda. "Dark Matter: A Primer." *Advances in Astronomy* (2011): http://dx.doi.org/10.1155/2011/968283.

Gates, Evalyn. *Einstein's Telescope: The Hunt for Dark Matter and Dark Energy in the Universe*. New York: W. W. Norton, 2010.

Gott, J. Richard. *The Cosmic Web: Mysterious Architecture of the Universe*. Princeton, NJ: Princeton University Press, 2016.

Greene, Brian. *The Fabric of the Cosmos: Space, Time, and the Texture of Reality*. London: Penguin, 2005.

Gregory, Stephen, and Laird Thompson. "The Coma/A1367 Supercluster and Its Environs," *Astrophysical Journal* 222, no. 3 (1978): 784–99.

Guth, Alan. *The Inflationary Universe: Quest for a New Theory of Cosmic Origins*. New York: Vintage, 1998.

Hawking, Stephen W. *A Brief History of Time: From the Big Bang to Black Holes*. New York: Bantam, 1988.

Hirschfeld, Alan. *Parallax: The Race to Measure the Cosmos*. New York: Henry Holt, 2001.

Koestler, Arthur. *The Sleepwalkers: A History of Man's Changing Vision of the Universe.* London: Penguin, 1959.

Kolb, Edward. *Inner Space/Outer Space: The Interface between Cosmology and Particle Physics.* Chicago: University of Chicago Press, 1986.

Krauss, Lawrence, and Robert Scherrer. "The Return of a Static Universe and the End of Cosmology." *General Relativity and Gravitation* 39, no. 10 (2007): 1545–50.

Kristeller, Paul Oskar. *Renaissance Thought: The Classic, Scholastic, and Humanist Strains.* New York: Harper & Row, 1961.

Lattis, James M. *Between Copernicus and Galileo: Christopher Clavius and the Collapse of Ptolemaic Cosmology.* Chicago: University of Chicago Press, 1994.

Mahon, Basil. *The Man Who Changed Everything—The Life of James Clerk Maxwell.* Hoboken, NJ: Wiley, 2003.

Nicolson, Iain. *Dark Side of the Universe: Dark Matter, Dark Energy, and the Fate of the Cosmos.* Baltimore: Johns Hopkins University Press, 2007.

Omnès, Roland. *Understanding Quantum Mechanics.* Princeton, NJ: Princeton University Press, 1999.

Pais, Abraham. *Inward Bound: Of Matter and Forces in the Physical World.* Oxford: Oxford University Press, 1986.

Panek, Richard. *The 4 Percent Universe: Dark Matter, Dark Energy, and the Race to Discover the Rest of Reality.* Boston: Mariner Books, 2011.

Reston, James, Jr. *Galileo: A Life.* Washington, DC: Beard Books, 2000.

Ronan, Colin A. *Edmond Halley: Genius in Eclipse.* Garden City, NY: Doubleday, 1969.

Silk, Joseph. *The Big Bang.* 3rd ed. New York: Henry Holt, 2002.

Stephenson, Bruce. *The Music of the Heavens: Kepler's Harmonic Astronomy.* Princeton, NJ: Princeton University Press, 1994.

Stone, A. Douglas. *Einstein and the Quantum.* Princeton, NJ: Princeton University Press, 2013.

Tasker, Elizabeth. *The Planet Factory: Exoplanets and the Search for a Second Earth.* New York: Bloomsbury Sigma, 2017.

Thorne, Kip S. *Black Holes and Time Warps: Einstein's Outrageous Legacy.* New York: W. W. Norton, 1995.

Weinberg, Steven. *Dreams of a Final Theory: The Search for the Fundamental Laws of Nature.* London: Hutchinson Radius, 1993.

Weinberg, Steven. *The First Three Minutes.* New York: Basic Books, 1993.

INDEX

Alpher, Ralph, 100
Andromeda galaxy (nebula), 82, 161,
  167–68, 205–206
    Cepheid stars in, 78–79
    as galaxy, 79–80
anthropic principle, 242
antimatter, 117
    balance with matter, 63–64
    and charge symmetry, 65, 67
    discovery of, 60–62, 112
    domination of matter over, 63, 67,
      91
    energy released by, 61
    location of, 62–63
    production of excess, 65–66
antiparticles, 114. *See also* particles
astrologers and astrology, 14, 17, 19, 57
astronomers, 14, 168
astronomy, 14, 57
    radio, 190
    X-ray, 190
astrophotography, 53
atmosphere
    of Earth, 27, 74, 197, 204–205,
      232, 234
    life's requirement for, 234–35, 240
    of planets, 51, 128
    of a star, 194–95
    of the sun, 205
atomic nuclei, 31, 113. *See also* fusion
atoms
    absorption of radiation by, 109, 130
    behavior of, 241, 244

collapse of, 131, 219
helium, 128, 133
hydrogen, 128, 129, 133, 134, 139,
  192, 244
nature of, 69, 108, 113, 115
neutral, 110–11
primordial, 95, 98
repulsion of, 30
simple, 139
and spectral lines, 106, 109
See also recombination

baryogenesis, 63, 67
baryon acoustic oscillations, 164, 187
baryons, 71, 113, 117, 197
Bessel, Friedrich, 56–57, 74, 82
big bang model, 91, 94, 97, 101, 118,
  120, 128
biosphere, 225
blackbody radiation, 98–99, 101–102,
  106
black dwarf stars, 212
black holes
    at the end of the universe, 215, 216,
      217–18
    formation of, 125, 137, 154,
      256–57n5
    mass of, 148, 253n9, 257n2
    in the Milky Way, 137
    nature of, 137–38, 141, 212–13, 237
    relativity and, 172, 211
    supermassive, 137, 200
blazars, 199

INDEX

light-years, 74, 81, 166, 250n1 (ch. 5). *See also* parsecs
LINERs (low ionization nuclear emission-line regions), 199
LIRGs (luminous infra-red galaxies), 199
lithium, 71, 72, 92, 150
Local Bubble, 225
Local Group, 168–70, 207, 225, 230
local superclusters, 232
Lockyer, Joseph, 52
Lockyer, Norman, 46
logarithms, 19
luminosity, 76

MACHOs (MAssive Compact Halo Objects), 154
Magellanic Clouds, 74–75, 199
magnetic fields, 37, 115, 234
    self-reinforcing, 200
magnetism, 50, 52, 58
Mars, 23, 204
mass
    of black holes, 148, 253n9, 257n2
    electron, 118
    energy and, 85, 86, 111, 116, 132, 193
    of galaxies, 144–45, 153–54
    of stars, 196
    of the universe, 151
matter
    and antimatter, 62–64, 92, 117
    density of, 128, 183–84
    excess of, 65
    in the Milky Way, 61
Maxwell, James Clerk, 50, 110
Mercury, 23, 148–49, 204
    orbit of, 23, 204, 248–49
mesons, 113

Messier, Charles, 49–50
microwave antenna, 100–101
Milky Way galaxy
    appearance of, 129
    black holes in, 137
    collision with Andromeda, 205–207
    evolution of, 137, 203, 212
    formation of, 151
    as galaxy, 225, 228–29
    images of, 166–69, 190
    intelligent species in, 240
    and the laws of physics, 244
    location of a Cepheid star in, 77
    matter in, 61
    observation of, 190
    as one galaxy among many, 225
    orbits within, 190
    photo, 229
    planets in, 228
    size of, 78–79, 81–82, 160, 199, 200, 226
    stars in, 95, 160
Minkowski, Hermann, 84. *See also* space-time
MOND (Modified Newtonian Dynamics), 152–53
monopole, 37–38, 39
moon (Earth's), 15, 26, 43, 54, 166, 204, 234
    orbit of, 44, 47
    recession of, 47
    sketches of, 21
moons
    of Jupiter, 26
    of the outer worlds, 43, 53, 204, 230, 234, 239
    of Saturn, 26
Mount Wilson Observatory, 78, 143

INDEX